Seeing the Southern Sky

PRACTICAL GUIDES SERIES

Seeing the Southern Sky

DAVID CALDER

WHITCOULLS PUBLISHERS
Christchurch Sydney London

Acknowledgements

I acknowledge with gratitude the assistance of Alan Gilmore and Pam Kilmartin (Mrs Gilmore), recently of the staff of Carter Observatory, who checked my manuscript and drew the charts; the Carter Observatory Board and Observatory Director, Dr Murray Lewis, for making the charts available; the California Institute of Technology, the Royal Astronomical Society (London), Mount Palomar, Mount Wilson, Radcliffe and Cape of Good Hope observatories for the use of photographs; Mr W. J. H. Fisher and the late Mr I. L. Thomsen, former directors of Carter Observatory, and Mr G. A. Eiby, former editor of the Royal Astronomical Society of New Zealand, whose writings gave me access to material not found elswhere; my son Graeme who contributed some calculations; my daughter Susan who held copy; and my wife Barbara whose efficient management of our home gave me the time to write the book.

First published 1977

Published and printed by
Whitcoulls Limited
Christchurch, New Zealand

ISBN 0 7233 0490 4

CONTENTS

Cover Picture

PLEIADES CLUSTER. Matariki of the Maori and the Seven Sisters of northern myth, photographed by the 1.2 m Schmidt camera at Mount Palomar Observatory. The stars visible to the unaided eye or in a small instrument are blue giants (the large ones in the photo). Between them, many hundreds of lesser stars and the remnants of the nebulosity from which they all condensed add to the strangely shimmering appearance of this beautiful group. The cluster is in the constellation Taurus and can be found in the northern part of our sky during the summer.

FOREWORD

It is with great pleasure that I write a foreword to this book.

Astronomy today is advancing rapidly, but advancing in a way that makes it hard for the average 'man in the street' to follow. Quasars, pulsars, X-ray sources, to name just a few, are not easy going and understanding astrophysics is a hard task if one has no background whatever.

Even the study of the planets by spacecraft brings with it some pretty fearful concepts and calculations.

This book fills the gap and introduces the novice to his environment, the sky and some of its wonders. What makes this book important is the fact that it is written by a New Zealander for New Zealanders and presents the sky as we see it, not as it is seen in the Northern Hemisphere as is the case with most books, great as some of them are.

David Calder has been known to me for years as a very able astronomer. Currently he is Publicity Officer for the Royal Astronomical Society of New Zealand—need I say more?

PETER READ, F.R.A.S.

INTRODUCTION

The Oldest Science

Astronomy is the oldest science. It was born when men first looked at the night sky, wondered, and began to deduce facts that are not obvious at first glance.

The stars gave mankind the calendar, a navigation system and the beginnings of mathematics. The night sky was the trigger of the inquiring factor in the human mind. It has been said that if life evolved on the surface of a planet such as Venus, with the sky forever hidden by a veil of cloud, no science and little intelligence would ever develop there.

Neither chemistry nor dynamics, botany nor mathematics would appear where the stars could not be seen. The largest numbers ever used in human calculations have appeared twice, several hundred years apart: once when Inca priests projected astronomical predictions aeons into the future looking for significant astronomical events; again when modern scientists first calculated the total number of electrons in the universe. Here were ancient astrologer-priests and modern astronomers with one thing in common: the star-born gift of mathematics.

Astronomy, by definition, is the science not merely of everything beyond the Earth, but also of everything that pertains to Earth as a planet. It concerns itself equally with sub-atomic particles in a rarefied gas, with the total mass of the universe, and with everything that lies between. Yet, one of the most marvellous things about it is that it can be studied at almost any level.

It can be a casual hobby, picked up and set down at will, an absorbing passion, or a demanding profession. It can be pursued at little cost or at great expense—and it is almost entirely safe. Almost, but not completely: the observer who studies the Sun is confronting the most dangerous object within light-years of Earth. Study the precautions on page 65 and follow them carefully. An astronomer can do without almost anything except eyesight, and the Sun can take that in an instant.

PART 1

The Southern Sky

The southern sky stretches over half of the planet Earth and is visible from more than half its surface, yet it remains the least-known part of the celestial sphere that surrounds us. This is all the more remarkable because the southern sky is by far the richer, by far the more rewarding of the two halves.

This was warmly attested by the Dutch-born astronomer Dr Bart J. Bok, formerly Director of Mt Stromlo Observatory in Australia, when he commenced an address before the august Royal Astronomical Society in London with the words, 'Gentlemen, you live under the wrong half of the sky.'

Indeed they do, but we in the south live under the right half and it is the purpose of this book to introduce all those interested in the universe around us to the striking, fascinating, brilliant southern sky visible from New Zealand.

This is not a textbook on astronomy. It is a guide to what the reader can easily see in the night sky, and also a bridge between other books and the sky. Although the situation is much better today than it was a few years ago, too many books concentrate on the northern sky to the almost total exclusion of the south. They do not ignore southern objects, for far too many are far too important, but they omit to tell the reader how to find them.

It is the purpose of this book, therefore, to tell you not merely that Alpha Centauri is the nearest visible star, but also that it is one of the Pointers to the Southern Cross; that the two Pointers are brighter stars than the famous northern Twins, Castor and Pollux, and that both pairs may be clearly seen from the south; to stress that the brilliant star clouds near the heart of our galaxy, which barely rise above the English horizon, pass almost overhead in the southern winter; to show you the two brightest and closest of the magnificent globular clusters, not the pale ghosts with which northern viewers must content themselves, and to guide you to the Magellanic Clouds, those two great star families which lie far beyond the boundaries of our own star system yet are

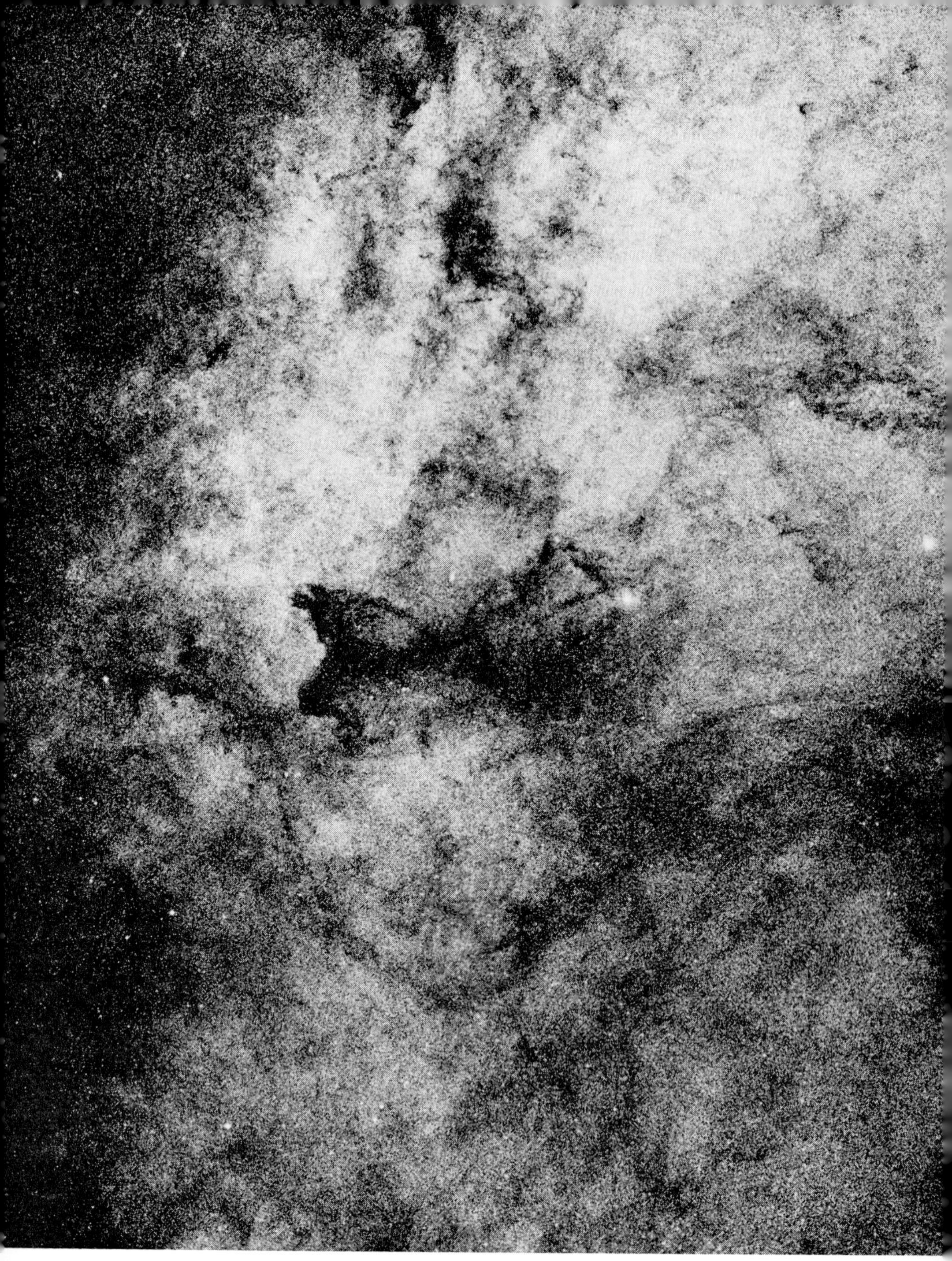

satellites to it, two of the most intriguing and intensively studied objects in astronomy today.

It might be assumed that the planets, at least, divide their favours evenly between the two hemispheres of Earth and for the most part they do, but both Mercury and Mars are better observed by viewers who live in the south.

The real tragedy is that the great observatories were built where the money was, and that was mainly in the north. For many years, of course, the resources of the northern sky were entirely adequate, but as astronomers plunge deeper into the mysteries of the universe they must turn increasingly to the best representatives of each class of object in the sky, be it the largest, closest, brightest, or unique, and a surprisingly large proportion of these are in the southern sky.

That is why big new telescopes are being installed in observatories in Australia (Mount Stromlo), South America and other southern countries, and why New Zealand, for long an astronomical wilderness, at last appears to be on the verge of taking part in that contribution to world astronomy that its position uniquely equips it to do.

Astronomy, like any science, is a search for answers to questions. Hopefully this book will give you some of the background to those questions, and show in what parts of the sky the answers are being sought.

STAR CLOUD IN REGION OF SAGITTARIUS. Not glowing gas but stars clustered thickly toward the centre of our galaxy. This star field, photographed by Mount Palomar's 1.2 m Schmidt camera, gives a good idea of the high density of stars in the central regions of the galaxy. The Southern Hemisphere in winter looks directly toward this rich region, which we see as the brightest part of the Milky Way. It is one of the principal reasons for astronomers' increasing interest in the southern sky. *(Hale Observatories)*

Chapter One

HOW TO LOOK

It is not necessary to be an astronomer to grasp the broad outline of the universe as revealed in the night sky, nor is it necessary to have expensive equipment. No instrument ever built outranks the human eye and the perceptive brain behind it, although equipment undoubtedly enhances the eye.

The sky on a clear, moonless night is impressive and beautiful. It may also be overwhelming if an observer looks at the whole and tries to make sense of it. If you looked at the map of a strange city and tried to grasp it all at once, you would be just as confused. What has to be done is to identify the observer's position on the map, then work carefully outward from one clearly identifiable landmark to another. This is precisely what this book will help you do with the night sky.

Actually, it is not as bad as it looks; it always comes as a surprise to learn for the first time that only some 6000 stars are individually visible to the unaided eye, and that something less than half this number will be visible to an observer on any given night. There are, of course, a few objects which are not stars, but they are indeed few—a handful of planets, some glowing and some dark patches of gas and dust and the soft glow of star clouds too distant for individual stars to be seen.

You don't need a telescope. This book is based on the assumption that most readers will be using their eyes alone, and that some will have binoculars. If a telescope seems inseparable from astronomy, think of it this way: a botanist needs a microscope if he is going to study pollen grains or cell structure, but the whole flower is more easily seen, and more beautiful. For the most part, we will be looking at whole structures rather than component parts.

Buy Carefully

If you feel that you must have a telescope, don't buy anything bigger than a 75 mm refractor until you have thoroughly exploited its potential and know that you have passed far beyond the scope of this book. It is deceptively easy to buy a telescope which is beyond a beginner's ability to use properly and this can be a costly mistake. Ideally, if you must have an instrument, buy a good pair of binoculars. Telescopes are expensive and specialised; binoculars far cheaper and more versatile. A family budget that may not stretch to the cost of a telescope may well manage a pair of binoculars that will serve the deer-stalking, bird-watching, horse-racing, you-name-it activities of several people in addition to the budding astronomer.

Binoculars will show you something of everything, and a few things much better than a telescope can. No observer has ever seen the full breadth of the Pleiades Cluster or the glory of the star clouds at the heart of our galaxy in the narrow field of a telescope. They are magnificent in wide-field binoculars.

A word of warning: 'power' is a largely misunderstood term, and in much advertising is downright misleading. It is a technical word, and it does not mean 'powerful'. If 'magnification' is substituted for 'power', we can discuss it with less danger of being misled.

Detail: Objective and Magnification

What really matters is the size of the 'objective', the big lens at the top end of a refracting telescope or binoculars. In a reflecting telescope the objective is a mirror. The bigger the objective, the more light the instrument will gather, and the better it will 'see'. For example, a double star just showing as an elongated image which no amount of magnification will separate in a 63.5 mm telescope can, in a 75 mm instrument, be clearly seen as two separate stars. Similarly, two stars even closer will form a single image in the 75 mm telescope but a 100 mm instrument will separate them, and so on.

Actually it isn't the diameter that matters, but the area, and the area increases with increasing diameter much more quickly than might be expected. Have you ever smiled at the Russians for the one-upmanship of building a 6-metre telescope, just 1 metre bigger than the 5-metre giant on Mount Palomar? Try working out the areas of circles of 5 metre and 6 metre diameter some time!

Magnification arises in the eyepiece and may have any value, from seven in some binoculars to several hundred in a large telescope. It may be changed in a telescope by changing the eyepiece. Increased magnification means a bigger image, but the easy assumption that 'bigger is better' could hardly be more wrong. In the first place, increasing the magnification shows a smaller area of sky. Secondly, although part of the image is now bigger, it is also fainter: the same amount of light has been spread over a greater area. Thirdly, increasing the image size does not make more detail visible, it only makes larger what was visible before.

Most importantly, the higher the magnification, the more difficult it is to hold the instrument steady. This is another, and possibly the best, reason for staying with a low magnification. In any case, the higher the magnification, the faster an object drifts out of the field of view of an undriven instrument.

There are certain optimum combinations of magnification and objective diameter which yield brilliant images in a wide field. The binoculars described as '7 × 50' (seven magnifications of the image produced by 50 mm diameter objective lenses) are a case in point. These binoculars, sometimes called 'night glasses', are the best all-around instrument and ideal for astronomy.

You will learn a great deal from trying out a few pairs in the street outside a retailer's shop. If you compare, say, 7 × 50 and 16 × 50 instruments, you will readily notice that the field in the latter is much narrower, that they are much harder to hold steady, and that the image is fainter.

A point to remember: binoculars show the view right way up, as the unaided eye sees it. So does a telescope intended for terrestrial work. This 'right way up' view, however, requires an extra lens which is not fitted to a telescope intended only for astronomical work. An astronomical telescope, therefore, shows an inverted, or 'upside-down', view. This normally matters only in respect of the Moon or in a few other cases in which the observer may notice the difference between what is seen with the unaided eye or binoculars, and the view in an astronomical telescope.

Mounting

Because the magnification used in a telescope is much higher than in binoculars, it must be mounted so that it is completely rigid and free of vibration. This requires a mounting much more robust than is needed to support the weight alone; the optical path will not tolerate any flexure or 'give', in the mounting at all. Many small telescopes of quite good quality are sold with woefully light mountings, frequently so light that the telescope cannot be effectively used on them for astronomical purposes (far less demanding terrestrial work may be possible).

Nothing even remotely like a photographic tripod, even a good, strong professional model, is sufficient. Most good mountings are one of two types: a rigid pillar permanently fastened to a large concrete block; or a very strong, heavy tripod. In the latter case, you will be quite happy to carry the tripod out by itself and make a second trip with the telescope. If the tripod can be comfortably carried in one hand, it's far too light.

Quality and Cost

Excellent instruments are made in Japan, Europe and America. The least expensive are not necessarily inferior, but some rubbish is produced in all countries. Beware of extravagant advertising claims, and any talk of 'high power' as if it was a good thing. Finally, some books (and advertising) give the impression that an equatorial mounting and drive are essential. For our purpose, such things would be an expensive waste of money, so it doesn't matter in the least if you haven't the faintest idea what an 'equatorial' is.

I have never owned a telescope, but do own good 7 × 50 binoculars. For the several years that I had charge of the Gisborne Astronomical Society's 125 mm refractor, I used it on a heavy tripod mounting. I have no choice but to use the equatorial mounting of Ward (Wanganui) Observatory's 24 cm refractor, but there are times when it is not the most convenient way to operate the instrument.

If you plan to spend more than the price of binoculars, seek advice. You can get it from books, or more effectively from local or national astronomical societies. Don't hesitate to seek an opinion on the suitability of a make or model of instrument, or the condition of a second-hand one. Astronomical hardware tends to be expensive. Don't waste money by spending unwisely.

Refractors versus Reflectors

In the size of telescope that most non-professionals will buy, up to 75 mm, the refractor is superior in almost every respect. From 75 mm to about 150 mm the refractor is still slightly superior, but the reflector is so much less costly that, over 75 mm, it is well worth considering. Reflectors of 75 mm can be bought, but their efficiency is so low due to the large obstruction caused by the second mirror that they are not really worth considering.

The following comparison assumes equal aperture:

Reflector: Better colour performance; less expensive; eyepiece position often (though not always) more convenient; usually shorter; superior for photography; air current problems with open tube designs (closed tube designs avoid this); some obstruction of light by secondary mirror and supports; reflecting surfaces require periodic re-aluminising; optics may easily move out of alignment.

Refractor: Superior performance up to about 150 mm; more expensive; closed tube avoids air currents and problems with dust on optics; optics not easily moved out of alignment; no obscuration of incoming light; maintenance and upkeep minimal; colour effects unavoidable; tube usually longer; eyepiece awkwardly placed when viewing at high altitude angles.

In general, reflectors tend to be of shorter focal length (particularly when of the most common, Newtonian design) and, in addition to superior colour performance, give the best views of nebulae and clusters. Refractors tend to be of longer focal length and give the best views of moon and planets.

The best choice in a small instrument is a 75 mm refractor. If buying a larger instrument, serious consideration should be given to a reflector with special reference to the kind of observations it is intended to make with it. Consideration should also be given to building a reflector (see Appendix Four).

The 'Seeing'

The first thing you will find out when you come to look at the sky, with or without instruments, is that it is never as simple as 'clear' or 'cloudy'. It isn't the ability to see that counts, it's the quality. This depends on both the steadiness and the transparency of the air. Looking straight up at the zenith (the point in the sky directly overhead) you are looking through about 16 km of air (although traces persist out to a much greater distance) and usually only the lower 1500 m matters. When you look at an object about 25 degrees above the horizon, the light path through the atmosphere has lengthened to almost 40 km, which matters a great deal. Below 25 degrees, transparency and 'seeing' are usually poor, especially in urban areas.

Even when the air is clear, movement affects 'seeing'—the greater the movement, the greater the effect. It may take one of two forms: the image may be reasonably clear and sharp but jumps around in the field as if the instrument were being shaken; or it may 'boil', as though you were looking at a molten object in violent turmoil. Some years ago, a New Zealand observer who knew just enough to get into trouble pointed a small telescope at a bright object one night in poor conditions and excitedly announced to his local newspaper that he had found an exploding star. It turned out to be the planet Mars. Generally the bigger the instrument and the higher the magnification you are using, the worse the effects will be (another excellent reason in favour of small instruments and low magnification). An unsteady image is usually bearable, although fine detail and faint objects will be obscured, but the 'boiling' state is hopeless. Give up and wait for a better night. If the stars are visibly twinkling, seeing conditions will usually be poor.

Clouds, Dust, Smoke

All too often the sky isn't even apparently clear. Cloud is obviously present. It may cover all or most of the sky or be in isolated patches; it may be dense or veil thin. Even if the Moon and brighter stars can be seen through such murk, it isn't worth observing. Between cloud patches, however, conditions may be quite good.

Far harder to detect are the effects of high, thin cloud, dust, smoke (sometimes from Australian bushfires) and cirrus cloud (layers of ice crystals in the stratosphere). It is surprising how often a night which looks good to the unaided eye turns out to be hopeless through the telescope. The effect of haze, from any cause, is to reduce contrast in the image and thus make fine detail and faint objects harder to see. Ground lights and moonlight make the effects much worse.

The sky around the object you are viewing should be black. The extent to which it isn't is an index of the amount by which haze is sabotaging what you see. Borderline conditions may be bearable before the Moon rises, but not after.

INTERLUDE

A Practical Exercise

As the apparent rotation of the sky each day is only a reflection of the Earth's rotation, the sky appears to rotate around two fixed points directly above the north and south poles of Earth. Everybody knows about the North Celestial Pole, marked by nearby Polaris, the Pole Star, by which, in the Northern Hemisphere, it is so easy to obtain a true direction at night, cloud permitting.

It sometimes comes as a surprise to learn that the southern sky, too, has its pole, although one of the few things we lack in our otherwise admirably equipped hemisphere is a bright star to mark it. It is, however, quite easy to find and because knowing where it is will be quite important later in the book, now is as good a time as any to undertake a first practical task: finding the South Celestial Pole from two or more observations.

Go outside early in the evening and face south. Assuming that you have a reasonably open southern horizon, that you are not in the middle of blazing city lights and that the sky is reasonably clear, you should have little trouble finding the distinctive group of the Southern Cross. In January it will be low down, just slightly east of due south. Each night at the same time it will be slightly higher, reaching a position high overhead in June, then starting down the western half of its circuit to be back at the bottom in December. Refer to the star charts in the second part of the book if necessary, but if you can't identify the Cross, don't worry.

Now look carefully at the Cross, or if you haven't located it, any other distinctive group that takes your eye, and fix its position in your mind. Whatever group you select, if it is east of south, select a second group rather west of south and not too high. Any two groups east and west of the direction you are facing will do. Fix their position relative to each other and to the horizon in your mind, or, better still, make a sketch. Two hours or more later, go out and find them again. You should now have little difficulty in fixing the approximate position of the point around which they are apparently rotating—the South Celestial Pole.

Chapter Two

WHANAU MARAMA

Whanau Marama, the 'Children of Light', was the old Maori collective term for the celestial host and there is no doubt that those tohungas who made a particular study of the stars knew a great deal more than most people today of the daily and yearly movements of the sky.

It is perhaps a sobering thought that what we now plan to do is in some respects no more than a recapturing of the knowledge that for many centuries was the stock in trade of the priest and navigator among the Maori people, for no matter the extent to which their astronomy was embellished with astrology and myth, they knew celestial movements with sufficient certainty to make them serve two practical purposes: ocean navigation, and the calendar.

If we are to recapture some of that knowledge which was far more common in older times, we must make a start. Choose a clear night, go outside and look at the sky with eyes prepared to see. We will assume that we have been able to find a perfect situation, the summit of a low hill surrounded by almost level land stretching away to the horizon, with no bright ground lights to interfere with the seeing. Above stretches the inverted bowl of the sky, bright with the 'Children of Light'.

At first the view seems chaotic, but after a few moments a pattern begins to emerge. A few objects are distinctly brighter than the rest. Of these, perhaps two or three appear to burn steadily. They are probably planets which, in poor seeing, will also appear to twinkle. All the rest twinkle to some extent, the more so the closer they are to the horizon. The distribution is not even. The stars cluster thickest toward that band of misty light, the Milky Way, and are thinnest in those parts of the sky most distant from it.

If only a few stars are bright, very large numbers are faint. We can easily see that as the stars grow fainter their numbers increase sharply and it is hard to believe that if we counted them, they would number fewer than 3000. What this really proves is just how big a number 1,000,000 is (although Ministers of Finance are apparently not convinced). Here and there among the stars are objects, apparently little bigger than the star images, that are perceptibly fuzzy. Other hazy objects have appreciable size, and look like pieces that have broken off the Milky Way.

We are looking outward across a vista that dwarfs the Earth and confounds the imagination if we try to think in terms of actual distances. This, however, we will not do. It is sufficient for our purpose to know the relative scale, that the stars are immensely more distant than the planets and that even these distances are dwarfed by the vast gulfs that separate the stars from some of the objects that we can see quite plainly far beyond them.

With the Eye Alone

At this point, the question most likely to come to mind is, how far can we see with the unaided eye?

With an unobstructed southern horizon, on almost any clear night in the Southern Hemisphere—but most especially in spring, summer and autumn—it is possible to see quite plainly the two Clouds of Magellan, objects from which the light, leaping across space at almost 300,000 km/sec, has taken 160,000 years to reach us. These objects, then, are said to be 160,000 light-years away.

The foreground of this vista is the realm of six planets, all at some time visible to the unaided eye: Mercury, Venus, Earth and its Moon, Mars, Jupiter and Saturn. They are spread across a radius of the solar system measured in hundreds of millions of kilometres, but beyond these six lie another three, Uranus, Neptune and Pluto, which extend that radius almost four times further.

From our hilltop we can see the Southern Cross and the two bright Pointers, low down to the south if it is summer, high above us in winter. The Pointer most distant from the Cross is the nearest visible star, yet it lies distant from us across a gap that cannot be measured even in terms of the radius of the solar system. Its light takes three years four months (3 1/3 light-years) to reach us. The other Pointer, nearest the Cross, is 300 light-years away. From many of the stars we can see, the light has been thousands of years upon its journey.

The Galactic System

All these stars, everything we have mentioned except the Clouds of Magellan, belong to a single great family, a star system called a 'galaxy' of which the Sun is a member. We can see only a few of the millions of stars of which it is composed, and deduce the existence of many more in the soft glow of the Milky Way, but between and beyond these are millions more—stars beyond stars, too faint and distant for our eyes to see, stars hidden behind the dense clustering of others, stars seemingly without end and without number.

Yet even this huge accumulation of stars is an infinitesimal part of the universe. Beyond it lie great gulfs and beyond those again more galaxies—the Magellanic Clouds, closest of all, and then more and more, in numbers that overwhelm the numbers of stars in our galaxy. They march outward to the edge of the universe all around us, galaxies large and small, some like ours and some most unlike it, the building blocks from which the material universe is constructed.

To become familiar with this immensity of bright stars and black depths we must take it part by part, learning what each part is and where it fits into the overall scheme. When we have mastered all the parts, we can assemble them into a whole in the mind's eye, and have a complete picture of the marvellous universe in which we live.

We will be able to take some pride from the fact that we will then have a view of the universe more closely keyed to reality than that of many modern astronomers, few of whom ever look through a telescope or take any interest in what their unaided eyes can show them. They live in a world of esoteric photographs and computerised data.

The Changing Sky

If we extend our look at the sky to a number of nights in the year we will immediately notice changes. Basically these are easily explained. The sky appears to rotate once each day as Earth rotates, and to this is added one further rotation as Earth makes its annual journey round the Sun. Thus, to each daily rotation is added a small part of the yearly one, and each night at the same time we find that the sky has turned a little further to the west.

This is why we cannot see every object in the southern sky at all times of the year. If we are able to watch all night we can see about five-sixths of what the sky has to offer, but the remaining sixth is obscured by the light of the Sun. It is the extra rotation of the sky added by the Earth's yearly orbital journey around the Sun that slowly changes the part available to us during the night.

The daily and yearly movements of the stars are thus simple and orderly, but the movements of the planets are more complex. Because of their great distances, the stars appear stationary relative to us (although they do have a motion of their own) and the only movement we detect is the reflection of Earth's movement. The planets, however, have a very noticeable movement. For those closest to the Sun and closest to the Earth, this movement can be considerable.

These apparently confused planetary movements bewildered ancient astronomers, who produced schemes of incredible complexity to explain them. After Copernicus had finally set the Sun and planets in their proper relative positions, however, the movements were easily explained. For us, the problem has a more prosaic form: star maps do not change from year to year. All that the observer need be concerned with is what part of the sky is visible on a given night. The movement of the planets, however, is not so simple. None, except Earth itself, moves in a cycle exactly a year long, so the planets are never in the same place on the same date in successive years. The observer, therefore, needs a source of current data to enable him to find the planets. Suggested sources are listed in the bibliography among the appendices.

INTERLUDE

Stones that Fall from the Sky

One of the best collections of minds ever assembled into a scientific society once received from a provincial district in a European country a stone said to have fallen from the sky. This miraculous event was attested by affidavits from the townspeople, and clearly the stone had been subjected to great heat—its surface was fused and pitted.

The learned men considered the stone and the affidavits, pondered, and announced their verdict:

'There are no stones in the sky. Stones, therefore, do not fall from the sky.'

The stone, no doubt, had been struck by lightning. Having thus set the science of meteoritics back by about 100 years, the learned men turned their attention to other things.

They should have been at Mokoia, $9\frac{1}{2}$ km south of Hawera in South Taranaki, just after noon on 26 November 1908. Accompanied by a thunderous explosion and a brilliant flash of light, a shower of incandescent stone fragments fell to earth, the main mass of the meteor streaking away to the south to fall into the sea off Wanganui.

The largest of the fragments recovered at Mokoia is now on display in Wanganui Regional Museum, a dark slab of ordinary-looking stone, with only its fused and pitted surface to show that once it came storming down the sky at the heart of an incandescent fireball.

Go outside on a clear night and sit where you can comfortably watch an area of sky more than halfway up to the zenith (the point in the sky immediately overhead) for an hour. Count how many streaks of meteor trails you see in this time. Most of them will be no bigger than a grain of sand, and 100 km up. The occasional brighter one may be slightly bigger, slightly lower.

Averaging out of meteor sightings yields a 'sporadic rate' for these random objects of one every seven minutes for such an observer. You may see far fewer, or far more.

Chapter Three

THE SOLAR SYSTEM

We will begin our study of the component parts of the universe with the solar system. It might be more logical to begin with the broad sweep of the galaxies and proceed downwards to smaller and smaller parts, but the solar system is where we are, so it is there that we will start. In fact, our view of the universe is closely bound to our position in it and the planetary observing platform from which we look out at it, so there is equal logic in beginning our voyage of exploration from Earth.

The nine planets which orbit the Sun fall neatly into four groups: four inner, Earth-like planets (Mercury, Venus, Earth, Mars); Jupiter, a lonely giant, unlike any other planet; three lesser giants (Saturn, Uranus, Neptune); and Pluto.

The four inner planets are all what we would expect a planet to be on the basis of our experience of Earth. Two are smaller than Earth, two are closer to the Sun than Earth and one is more distant. The only differences between them are the differences we would expect to follow from size and position. Thus, Mercury is much smaller than Earth and much closer to the Sun, so it lacks an atmosphere and is much hotter. Venus is much the same size as Earth, but closer to the Sun. It has an atmosphere and is hotter. Mars is smaller and further from the Sun. It has less atmosphere and is much colder.

Beyond Mars, nothing in our experience equips us to recognise four huge bodies and one tiny one as planets having anything in common with Earth.

Jupiter, the giant, is bigger than all the other planets put together. Unlike any other planet, it radiates to space far more energy than it receives from the Sun. Later we will discuss a theory that it almost became a star.

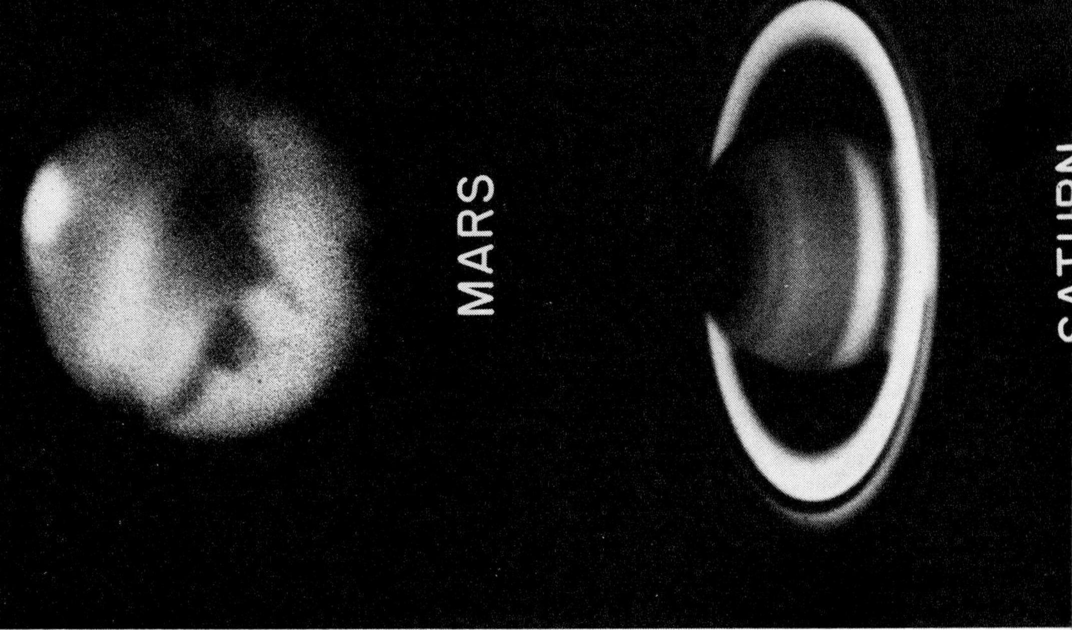
MARS
SATURN

Saturn, Uranus and Neptune are also big, but not so big as Jupiter. They are huge compared with Earth, and cold, solid cores lie hidden below deep, poisonous atmospheres.

Last is Pluto, small, probably like Mercury, but so distant that its size is not easily measured. It swings slowly around the Sun in the grip of a cold more intense than anything we can imagine on Earth.

These nine planets, with the Sun, are the major elements of the solar system, but they are not the only members. Between the orbits of Mars and Jupiter orbit a host of tiny bodies called asteroids, or minor planets. A few swing far in towards the Sun, even inside the orbit of Mercury, and a few sweep far out towards Saturn, but most are fairly strictly confined to their zone between Mars and Jupiter.

Not unlike the asteroids, but mainly smaller and with orbits of all sizes distributed around the Sun in all directions, are the meteors, shapeless pieces of rock or iron ranging from thousands of tons down to grains of dust. The former originate from asteroids, the latter mostly from comets. It is these which occasionally plunge into Earth's atmosphere, clearly visible as streaks of light high in the night sky. Very occasionally, one of the rare large pieces survives its fiery plunge long enough to reach the surface of the Earth. Meteorites were the only pieces of extraterrestrial material available for study until the first men returned from the Moon.

FOUR PLANETS. Even when taken with the largest telescopes, photographs characteristically show only as much detail as the visual observer can see with a moderate instrument. Mars, Jupiter and Saturn were photographed with the 2.5 m Mount Wilson telescope. The observer with a 100 mm refractor or 152 mm reflector would see essentially the same detail in good conditions. Pluto was photographed by the 5 m telescope on Mount Palomar. It is too faint to be seen in a small instrument. The four separate photographs have not been printed to the same scale. If to the same scale, Jupiter would appear much larger than Saturn and Mars much smaller. *(Hale Observatories)*

Heralds of Doom

Also distributed randomly around the Sun are large numbers of comets, of which the largest and brightest were regarded until almost modern times as the heralds of a doom about to descend on Earth. They are the least substantial but most spectacular of the Sun's family and, while some complete their orbits quite close to the Sun, others retreat to distances far greater than Pluto. They remain among the least understood of celestial bodies.

The orbits of the planets all lie close to the plane of the Sun's equator, so the apparent tracks of the planets across the sky all lie within a comparatively narrow band (18° wide) called the 'Zodiac'. Through this runs the 'ecliptic', the projection of the yearly path of the Sun against the celestial sphere and also, therefore, the average path of the Earth and other planets. The constellations through which the Zodiac passes are known as the 'constellations of the Zodiac' to which astrology attaches great significance. It is worth noting, however, that astrology recognises only twelve such constellations, although there are in fact thirteen —Ophiuchus also qualifies. A further point of interest is that, for more than 2000 years, astrology has ignored the slow shift in the sky caused by the 'wobble' in the Earth's axis of rotation, so that there is now a discrepancy of 30 degrees, a whole constellation, between the astrological tables and astronomical reality.

Chapter Four

THE INNER PLANETS

The First and Smallest

We will start our astronomical journey with Mercury, the smallest planet and the closest to the Sun. First, here are the 'vital statistics' to make comparisons easier: equatorial diameter, 4880 km; mass, 0.055 of Earth's; distance from the Sun, 57,910,000 km; sidereal rotation (the planet's own day), 58.6 Earth days; revolution (year), 88 Earth days. It was only recently discovered that the planet's sidereal rotation period is only two-thirds of its year. It was previously thought that year and day were equal, the planet turning one face eternally to the Sun.

The new value for the day makes this tiny planet much less bizarre than it was previously thought to be, but it would still be a most uncomfortable place to visit. For almost 30 Earth days, the Sun is wholly above the horizon, and it is a sun such as we have never seen from Earth. It looms three times larger in the black, airless sky of the planet and pours upon its surface nine times the radiant energy that reaches Earth. Nor is this great energy flow the almost benign sunlight we know on Earth—it is the raw, undiluted, unfiltered output of the Sun, all the way from deadly ultraviolet and x-rays to the shattered debris of atoms that form the solid particles of the 'solar wind'.

Under this awful heat, rocks may reach red heat in the hottest part of the day and metals such as lead and tin, if they exist, flow like water. After sunset, the heat-wracked surface enters a thirty-day night and temperatures begin a long drop to very low levels, checked only by what little warmth is stored in the surface rocks.

Because of the small radius of its orbit, Mercury is never far from the Sun in the sky and, depending on which side of the Sun it happens to be at a given time, is visible only briefly before the

Sun rises or after it sets. It can never be seen in a completely dark sky, only in twilight, and so is rather difficult to study. It is, however, comparatively easy to identify as a bright point of light. Mercury has a reputation for being elusive (it is said that the great Copernicus never saw it) but Mercury's orbit places it further from the Sun in the southern sky and New Zealand weather is usually good enough for long periods not to interfere too greatly with objects near the horizon. When you have found it, try to pick it up again on a number of successive nights and note how rapidly it moves.

Mercury is a small, curious world and one that may still have surprises in store for astronomers. Space probe pictures show a surface cratered like the Moon, although sharp features have been blurred by the effects of thermal erosion as rocks crack and split in the extremes of temperature. Mercury has no moon.

A Twin to Earth

Venus is 108,210,000 km from the Sun. Its mass is 0.8150 of Earth's and its diameter 12,100 km. It rotates on its axis in 243 Earth days and its year is 224.7 Earth days. The planet has no moon.

In size, Venus is almost a twin to Earth, but in other ways is like no other planet in the system. Note that its day is longer than its year. It also rotates about its axis in the opposite direction to all the other planets, behaviour for which no explanation has yet been found.

If Mercury is elusive, Venus frequently thrusts itself upon our attention with a brilliance that makes it at times the third brightest object in the sky, after Sun and Moon. Although, like Mercury, it is closely bound to the Sun by an orbit smaller than Earth's, it is often far enough from the Sun to appear in a completely dark sky after sunset or before sunrise. At its brightest, it throws a clearly visible shadow (on Moonless nights, away from ground lights) if the observer thinks to look for it.

Like Earth, Venus is surrounded by an atmosphere, but there the similarity ends. The planet's surface is hidden below a dense veil of cloud that has never lifted (although occasional markings believed to be on the surface have been glimpsed). These brilliant white clouds are the main reason why the planet is so bright in our sky.

The heat engine of the planet's weather is powered by solar radiation twice as intense as that which reaches Earth. Space probes have studied the atmosphere in ultraviolet light and find it in violent motion, although the weather patterns appear to be highly organised. Early attempts to land instrument packages on the surface were largely unsuccessful, bare confirmation of predicted high temperatures and pressures being radioed back before the instruments succumbed after a few minutes' exposure to the hostile conditions.

Russian Venus Probes

Late in October 1975 the Russian probes Venus Nine and Ten both landed instrument packages which transmitted pictures back to Earth. Total data transmission time from inside the atmosphere for the two vehicles was about 2 hours, including more than 50 minutes of transmission from the surface. Both vehicles then ceased transmission. The transmitted data confirmed that obtained earlier and considerably expanded it.

The pictures from the first vehicle showed a landscape very different in appearance from what had been predicted. In place of the barren, windswept landscape of dunes expected, the Russian scientists saw a landscape littered with boulders, ranging up to 1 m in diameter. Many of the edges appeared to be sharp, negating the predictions of a surface long worn smooth by severe wind and thermal erosion.

The second vehicle, landing more than 1000 km away, found a significantly different landscape, smoother and looking like the bed of an ocean or river. The Russians also reported strong evidence of volcanic activity on the surface. Another prediction not confirmed was certain optical effects expected from the gases under high pressure at the surface: long objects, flat planes and the horizon did not, apparently, bend upward as a result of refraction of the light. There was some suggestion that this was inconsistent with the confirmed pressures.

Atmospheric temperature and pressure at the surface are

undoubtedly high. Conditions are probably also highly corrosive. The dominant gas appears to be carbon dioxide with little free oxygen or water vapour. The presence of the carbon dioxide is itself a puzzle—it should break down in the fierce sunlight, yet it is apparently stable. One effect of the carbon dioxide is to increase the surface temperature. The gas is transparent to the incoming short-wave-length solar radiation, but opaque to the long-wave-lengths (infrared) re-radiated from the surface. This heat, unable to escape freely, contributes to a rise in temperature. This heat trap mechanism is known as the 'greenhouse effect' because it takes place also in glasshouses where plants significantly increase the carbon dioxide concentration in the air.

The instruments landed on Venus (and those which have made close approaches to Jupiter) do confirm one general prediction—that much remains to be discovered about the planets of the solar system. The study of the solar system has tended to proceed in a series of waves of intense interest in the planets, interspersed with periods in which planetary studies have had a low priority with astronomers. We are now on the crest of a great new wave of exploration which bids fair to sweep away much of what we believe now and reveal the planets as far more mysterious than they seemed when our closest view was through a telescope.

Seeing Venus

Venus, at times, approaches closer to Earth than any other planet. It is much like Earth in some ways, but clearly so hostile to life that man could not survive there without sophisticated equipment which is probably not yet even on the drawing-board.

When it first appears in the evening sky, Venus is close to the Sun and almost drowned in its radiance. It can then be as hard to find as Mercury, but night by night it draws away from the Sun until it appears in the twilight as a brilliant point of light visible long before the brightest stars begin to show. This increasing separation continues until it appears at its most brilliant in a dark sky, outranked only by Sun and Moon.

At this time, a telescope or (in good conditions) binoculars will show the planet looking like the Moon at first quarter. When first glimpsed near the Sun, Venus shows a near-full disc, but it appears

VENUS, photographed by the 5 m telescope on Mount Palomar. The planet is in a crescent phase after passing its period of greatest brilliance in the evening sky and beginning to swing between Earth and Sun. The brilliant, featureless top of its cloud-filled atmosphere shows no trace of surface markings in this photograph, although faint markings and indentations are sometimes seen along the terminator. *(Hale Observatories)*

small because the planet is then on the far side of its orbit. As it swings toward us, the diameter of the disc grows steadily bigger, but it changes phase as does the Moon, waning to a crescent as more and more of the dark side is presented toward Earth. After the time of greatest brilliance, the bright area shrinks rapidly as the planet swings between Earth and Sun, presenting only its dark side to us. After being too close to the Sun to be seen for a short time, Venus reappears in the morning sky and goes through the same sequence of changes in reverse.

For a long time around its period of greatest brightness, Venus can be found in broad daylight if the observer knows where—and how—to look. The best method is to estimate the separation between planet and Sun just after sunset (or before sunrise) when the planet is clearly visible in the twilight. Estimate the angle between planet and horizon (a handspan at arm's length is about 20 degrees), then add a bit more for the estimated distance of the Sun below the horizon. This angle can never be more than 48 degrees and is usually much less. Beware, however, of a tendency to overestimate angles low down in the sky and to underestimate high ones.

When the angle and direction of Venus from the Sun has been thus determined, use it to find the patch of sky containing the planet when the Sun is high. If you look carefully, you will find the brilliant point of light which is the planet clearly visible, although it is not always easy to pick up at first. It can be a little difficult to learn to focus the eyes at infinity while looking into the blue sky, but if the eyes are not so focussed, the planet will be out of focus and remain hidden. It helps if the Sun can be obscured behind a roof or tree.

When both at its brightest and high in the sky, the planet is easily found and many people lying on their backs sunbathing, gazing idly into the sky, pick it up. Few realise what it is—and another flying saucer story is born.

Venus's changes in brightness are great enough to be clearly discernible to the unaided eye if it is followed through the cycle from first appearance close to the Sun, out to maximum separation and back toward the Sun again. For part of the time with binoculars (and most of the cycle with a telescope) it can also be seen to change phase, as does the Moon.

From the time when it reaches greatest separation from the Sun in the evening sky, Venus takes about 144 days to swing back towards the Sun, cross to the other side, reappear in the morning sky and reach maximum separation on that side. Then the whole cycle repeats in reverse, taking about 144 days to regain maximum separation in the evening sky.

The Double Planet

The next object outward from the Sun is one of the most striking in the solar system, a double planet whose components swing for ever around their centre of mass. The larger has a deep atmosphere and extensive areas of water on its surface; the smaller is bare and barren, its surface scarred by the impact over aeons of meteor debris. We call them Earth and Moon.

It is only to us on the surface of the Earth that the Moon appears as a simple satellite in orbit around us. If we could observe Earth and Moon from one of the nearer planets we would clearly see that the Moon is no simple satellite, but a body of mass sufficiently large compared with Earth that 'double planet' is the only possible description. We will learn later that several satellites of the giant planets are larger than Earth's moon, but each is a much smaller fraction of the mass of its planet. To Earth and Moon belong the distinction of being the two most nearly matched, associated bodies in the solar system.

The double planet is 149,600,000 km from the Sun and its components are separated by 385,000 km. Earth is 12,756 km in diameter and the Moon 3476 km. The Moon is 0.0123 of Earth's mass. Earth rotates on its axis in 23 hr 56 min 4 sec (with respect to the stars); the Moon in 27.322 days. The pair revolve around the Sun in 365.256 days.

We will not devote too much time to Earth as most readers will have first-hand experience of it. It is worth noting, however, that one of the effects of the space programme has been to make us more conscious of Earth as a *planet* than man has ever been before. It is the prototype planet, the one by which we tend to judge the others (although this is really rather a self-centred attitude).

Our Companion

The Moon is small, but it is also close and therefore appears to us both big and bright—exactly as big as the Sun and second only to the Sun in brightness. We think of the Moon as being very bright, but this is largely illusion created by seeing it in a dark sky. Next time you see the Moon rising almost full just before sunset, try to position yourself so that you can see it alongside a stone wall or outcrop on a distant hill. It will be instantly apparent that the Moon is as bright as stone in full sunlight, and no brighter.

The Moon is a separate world, and always was, although it probably formed close to Earth and the two became linked very early in the history of the solar system. The Moon revolves around Earth (to take the simple view) in just over 27 days (relative to the stars), but appears to take $29\frac{1}{2}$ days, the period between two identical phases. These are the 'sidereal' (relative to the stars) and 'synodic' (relative to the observer on Earth) periods. For the Moon, the times of orbital rotation and axial revolution are equal, so one face is always toward the Earth. Note, however, that all parts of the Moon face the Sun at some time during its revolution, experiencing a 'day' of about 14 Earth days. During this long period of sunlight, temperatures rise very high, but as the Sun sets they plunge rapidly for there is no atmosphere to blanket the Moon and help it hold its heat.

Eclipse

The orbit of the Moon is not in the plane of Earth's equator, but lies at an angle which can vary from 18 to 28 degrees. Because of this, the Moon does not always pass directly between Earth and Sun at new moon. When it does, we see the result of the Moon appearing to be the same size as the Sun: for perhaps three minutes the Moon completely, but only just, obscures the Sun's brilliant disc and we see around the edge of the Moon the glowing outer atmosphere (corona) of the Sun. This effect, a solar eclipse, is one of the most beautiful and spectacular sights the universe can show us.

Sometimes, at full moon, the Moon passes through Earth's

shadow. The Moon is then said to be eclipsed. The Moon does not usually disappear in eclipse, for some light continues to reach it after being bent through Earth's atmosphere, but it takes on a ruddy, coppery hue and, strangely, appears obviously spherical, most unlike the flat face of the un-eclipsed Moon.

As it moves into and out of eclipse, the Moon passes through a zone of partial shadow (the penumbra) but this is not usually easily seen. The advance of the Moon into the total shadow zone (the umbra) is usually quite apparent and can be followed by the unaided eye. On rare occasions a lunar eclipse may reach one of two extremes—an eclipse so dark that the Moon disappears, or nearly so, or an eclipse so bright that the Moon does not appear to be dimmed, although a coppery hue is usually detectable.

There are at least three eclipses a year, sometimes more, and at least two are always eclipses of the Sun (though often only partial). We tend to think of eclipses of the Sun as being rare and those of the Moon as much more common, but this isn't so. Solar eclipses seem rare because they are seen only over a narrow track on Earth and it is necessary to be on that track to see the eclipse as total; the Sun appears partially eclipsed over a much wider path. Lunar eclipses, by contrast, are visible over a whole hemisphere of Earth, weather permitting, so far more people see them.

It is rare, indeed, just barely possible, for two solar eclipses to be visible from a single place within a single lifetime. Most people never see one. The last total solar eclipse visible in New Zealand was in 1965 (the narrow track crossed Northland) and there will not be another until 22 July 2028 (in Southland). On 16 January 1991 the track of an annular (not to be confused with 'annual') eclipse crosses central New Zealand. An annular eclipse occurs when the Moon is at its most distant from Earth and Earth is at its closest to the Sun. The Moon's apparent size is then slightly smaller than the Sun's, and a ring ('annulus') of unobscured Sun surrounds the Moon at mid-eclipse. In other words, the Sun is not covered completely and the eclipse is thus not total.

Whenever either a total eclipse or an annular eclipse is visible from some part of New Zealand, a partial eclipse is seen over the whole of the rest of the country. Just outside the central path the Sun is almost entirely obscured; in the most distant part of the country the Moon barely encroaches on the solar disc.

The Terminators and Viewing

At new phase, the Moon is close to the Sun in the sky and invisible in its glare. As it moves in its orbit, it draws away from the Sun and becomes visible as a narrow crescent, so thin and faint in the daylight sky that it is seldom seen. Twenty-four hours later, however, the waxing crescent is much brighter, much further from the Sun, and is easily found and followed from night to night as it grows to first quarter phase, when the disc is divided exactly in half. The edge of the sunlit area advancing across the Moon's face is the line of sunrise on the Moon. An observer standing on this sunrise line would see the Sun low in the lunar sky and, of course, it would throw long shadows.

Along the sunrise line (called the 'terminator') is the best place to look for the mountains of the Moon with binoculars or a small telescope, as the long shadows make them stand out in bold contrast. At first quarter, the terminator lies across three big formations that are particularly striking. They demonstrate the most obvious fact about the mountains of the Moon, that they are ring craters, ranging in size from hundreds of kilometres across (and plainly visible in small instruments) down to craterlets mere millimetres in diameter. A few ranges of more Earth-like mountains do exist.

After full moon, the sunset terminator begins its traverse of the disc, returning it to complete darkness at new moon. The first quarter moon rises at noon, full moon at sunset and last quarter moon at midnight.

A number of the larger formations can be identified easily in binoculars when near the terminator, and a few can even be seen with the unaided eye. It is a worthwhile exercise, using a good map of the Moon, to identify as many as possible. An observer who tries this will quickly find that the illumination of the lunar surface is seldom identical at the same time in successive lunar cycles. This is because the Moon 'rocks' both east and west, and north and south. These movements, called 'libration', do not concern us here, but the effect is easily detected.

COPERNICUS, photographed by Mount Wilson's 2.5 m telescope, is the finest ring-plain on the Moon. The walls are 90 km in diameter and the floor 64 km. The crest of the walls is 4000 m above the floor at its highest and nowhere drops much lower. Copernicus displays all the characteristics of a typical lunar crater: the outer slopes are gentle, the inner much steeper and are terraced; at first glance circular, it is actually a polygon made up of straight sections of wall; the floor is lower than the surrounding plain; relatively low central mountains rise from the floor; small crater pits appear on all parts of the main structure. The crater is younger than many lunar structures and retains both sharp outlines and bright appearance. It may be seen as a bright spot with the unaided eye at full moon, when it is also remarkable for the system of bright rays (visible in any instrument) radiating from it.

(Hale Observatories)

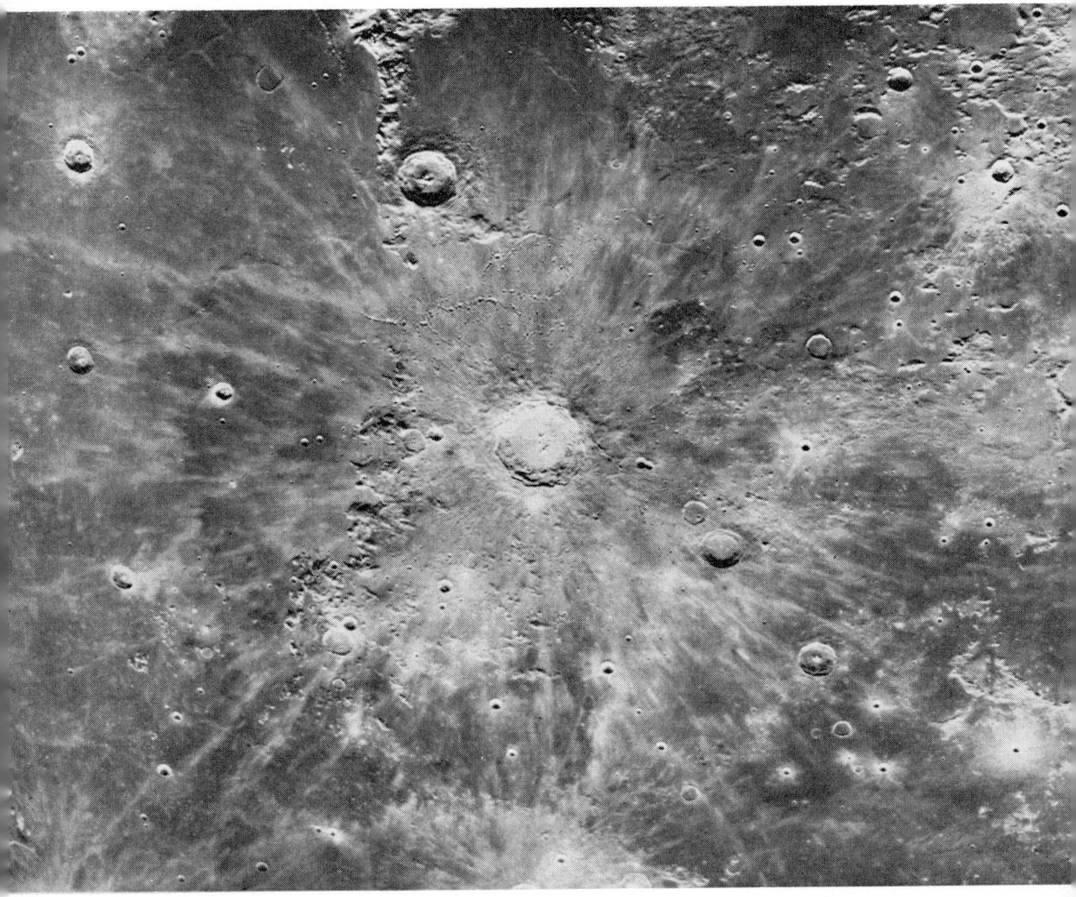

The Red Planet

The last of the inner planets is Mars, 227,940,000 km from the Sun. It is 6790 km in diameter and 0.1074 of Earth's mass. It rotates on its axis in 24 hr 37 min 23 sec and revolves around the Sun in 687 days. Note that its day is almost exactly the same length as Earth's, and it so happens that the inclination of the axes of the two planets is almost identical. Mars's year, however, is almost twice the length of Earth's.

Mars is the most Earth-like of the planets, though it is much smaller than the Earth and far enough from the Sun to be much colder. It has an atmosphere, clear for the most part so that we see the surface, but showing clear evidence of weather—great dust storms sweep the surface occasionally, obscuring large areas for weeks at a time.

If it is like Earth, it is like no part with which we are ordinarily familiar. Those like Sir Edmund Hillary who have been to the top of Mount Everest or to the South Pole have experienced conditions which are close to normal on Mars; only in such extreme areas could conditions on Earth nearly approach those in the temperate latitudes of Mars during the long Martian winter. This is especially true of the southern hemisphere of the planet, for the orbit of Mars is far from circular and during the southern winter the planet is much further from the Sun than at other times.

As the Martian autumn passes and winter begins, what little water vapour the atmosphere contains freezes out, appearing on the ground as a frost which grows more dense toward the pole. This, however, is only a signal of the coming winter. As the temperature continues to drop, carbon dioxide gas begins to freeze within the polar circle, building a glittering cap of 'dry ice' clearly visible from Earth in even a small telescope. The Sun shines small and distant in a clear, dark purple sky, giving only half the radiation Earth receives, and a thin bitter wind blows steadily across a landscape where only dust moves, locked in the freezing cold.

World That Died

There is some water, so life is at least possible, but Mars is probably a world that died before it had a chance to live, too small to hold the necessary protective blanket of a dense atmosphere or its internal heat, and now near frozen death. Across its surface blows the eternal wind, eroding and shaping it to an extent not seen on Earth. This is because on Mars there is not enough water to trap the fine particles of dust and reduce the effects of wind erosion. Cameras on spacecraft orbiting high above the surface have shown clearly the effects of this erosion. Crater walls are ridged and gullied as if by water and delta fans of debris spread from their base. It was once thought that this was almost certainly wind erosion, that it was highly unlikely that water had ever flowed there. In early July 1976, however, the close observations made of the surface prior to the Viking 1 and 2 landings suggested some free water did exist at the surface and that these markings might indeed be water erosion. These findings greatly enhanced the chances of finding life on Mars, most likely in a form akin to soil bacteria. The Viking landings were made as this book went to press.

If Mars is observed at intervals of a week or so after it appears in the evening sky and its position noted relative to the stars, it will be seen to move steadily westward for a while, then to stop, and then to move backward towards the east. After a while it halts again, then resumes its westward journey, having described a loop against the background of the fixed stars. This behaviour, characteristic of all the planets beyond Earth, is shown most markedly by Mars. It was this mysterious movement which so puzzled the ancients who assumed it was real, but in fact it is only a reflection of Earth's movement which, travelling much faster in its inside orbit, catches up to and then passes the slower outer planet.

Interesting Moons

The two moons of Mars are far too small to be seen in anything but a large telescope, but are interesting enough to be worth mentioning. They are called Deimos and Phobos (Flight and Fear, the legendary steeds of the God of War). Phobos is only 5950 km above the surface. It takes 7 hr 39 min to orbit the planet, racing up over the eastern horizon and across the sky, changing through half its phases as it goes, then disappearing in the west. Deimos is 20,100 km above the surface and takes 30 hr 17 min to complete one orbit. This is longer than the planet's day, though not by much, and Deimos remains above the horizon for long periods, going through its phases time after time as it creeps slowly across the sky. Both are small, and even from the surface of the planet would be faint and insignificant.

It is only fair to stress that Mars is a difficult, unrewarding object, even in a moderately large telescope. It is small, almost always distant, and a small instrument shows only the characteristic red colour, an occasional dark marking and, in season, a white polar cap. It cannot compare with the next two planets, Jupiter and Saturn, which, despite their much greater distance, are, because of their size, fascinating objects in even a small telescope. Jupiter will display its moons to a binocular observer.

INTERLUDE

Jonathan Swift and Lemuel Gulliver

Obtain a copy of Jonathan Swift's Gulliver's Travels *(a full edition, not the condensed version sold as a children's story), and find on about the fifth page of Chapter Three of Part Three, 'A Voyage to Laputa', the description given to Gulliver of the moons of Mars by the scientists of that country. Then read again the description of the moons of Mars in this book. Now reflect that Swift wrote* Gulliver's Travels *before 1726, and that Asaph Hall did not discover the moons of Mars until 1877.*

How did Swift know, more than 150 years in advance?

There was, in fact, neither miracle nor magic in Swift's prediction and the answer is quite simple: Swift was an educated man. Today that would not be enough but in the eighteenth century it was, because it was taken for granted then that to be 'educated' included an extensive knowledge of both astronomy and mathematics.

Swift wrote satire. All the lands that Gulliver visited were thinly disguised Englands, and the whole work was a clever, satirical criticism of England and things English. In Gulliver's visit to Laputa, Swift was exercising his wit on the infant Royal Society, whose members, despite doing much that was worth while, had managed on several occasions to make prize fools of themselves.

Swift was educated, and he was thorough. Knowing that the much more distant, but large, satellites of Jupiter had been discovered long before by Galileo, he sat down to find by calculation, assuming a reasonable advance in telescope efficiency, what manner of satellites Mars might have that were as yet undiscovered, but were discoverable.

He found that the limits imposed by the physical situation were so strict that only three possibilities existed: the planet might have no moon; it might have one; it could have two.

Taking the most interesting case, two moons, he analysed this hypothetical Mars/moons system in terms of Kepler's three laws of planetary motion. His calculations were impeccable, and indeed the only three possibilities that existed were the three he found.

More than 150 years later, Asaph Hall, working at Washington Observatory, discovered that the planet did indeed have two moons. Bound by the unchanging factors that had been apparent to an educated cleric a century and a half earlier, they behaved almost exactly as he had predicted they would.

People today are puzzled by the accuracy of the prediction only because, despite their more extensive knowledge, they are not in fact as well educated as was Jonathan Swift.

Chapter Five

THE ASTEROIDS AND THE LONELY GIANT

To reach Jupiter we must cross a gap much greater than that which separates the inner planets. It is not empty, but contains a host of small bodies orbiting the Sun: the asteroids.

The largest asteroids are only a few miles in diameter, and only approximately spherical. They can be detected in binoculars as faint points of light hurrying among the background stars. A few of these which plunge far in toward the Sun can pass quite close to Earth and it was the asteroid Eros which a former Astronomer Royal, Sir Harold Spencer Jones, used to refine the Earth-Sun distance before more modern methods made this one obsolete. Below the handful of large asteroids their size drops rapidly and they are not even approximately spherical, just large pieces of rock which grow steadily more numerous as their size decreases.

The asteroids number about 1800 down to the lower limit at which they can be detected (photographically; most are too faint ever to be seen) and there is no reason to doubt that the size sequence runs right down to particles a few millimetres in diameter. To many astronomers they are a nuisance, repeatedly forming images on photographic plates which have to be checked out in case they are something more important. The asteroid belt is sometimes referred to as a celestial gravel pit, but a few astronomers are keenly interested in these tiny members of the Sun's family.

Theories of their origin vary, but it is unlikely that they are the shattered remnants of a planet. Numerous as they are, their total estimated mass is that of a body smaller than Earth's moon. It is more likely that they represent primordial material which the nearby bulk of Jupiter prevented from condensing into a single body. They may be the 'genesis rock' of the solar system, the material from which the inner planets were made.

The Lonely Giant

Jupiter is truly a giant, bigger than all the other planets put together. It takes almost 12 years to swing once around the Sun, majestically carrying twelve satellites with it. The planet is 143,200 km in diameter, its mass is 317.89 times that of Earth, it is 778,340,000 km from the Sun and rotates on its axis once in just under 10 hours.

The effect of this rapid rotation can be quite clearly seen in a telescope, even a small one. The planet bulges quite perceptibly at its equator and the disc is distinctly elliptical in shape. For the observer with a small-to-moderate telescope, this is the most fascinating object in the solar system, and binoculars will reveal something of this dynamic world of constant change.

The twelve satellites range down to sizes so small that they have been photographed but never seen, and some move in orbits so distant from the planet that they are almost as much under the influence of the Sun as of Jupiter and their orbits take quite complex shapes.

The Galilean Moons

Four of the satellites, however, are big and famous; these are the objects which Galileo discovered when he first turned the newly-invented telescope to the night sky, and which so disturbed the established ideas of his age. Today we can walk in Galileo's footsteps, seeing these objects clearly, for even a modest pair of binoculars are a far more powerful optical aid than Galileo used.

Of these four Galilean satellites, three, Callisto, Ganymede and Io, are as big as Earth's moon, or bigger, and Europa is a little smaller. They, together with the biggest satellites of all the giant planets, can be regarded as worlds much like Earth's moon in most respects, except that because it is so much colder out in the realm of the giant planets, the biggest retain traces of atmosphere. The smaller satellites are irregularly shaped bodies, indistinguishable from asteroids, which is probably what they were before they ventured too close to one or other of the giants and were captured.

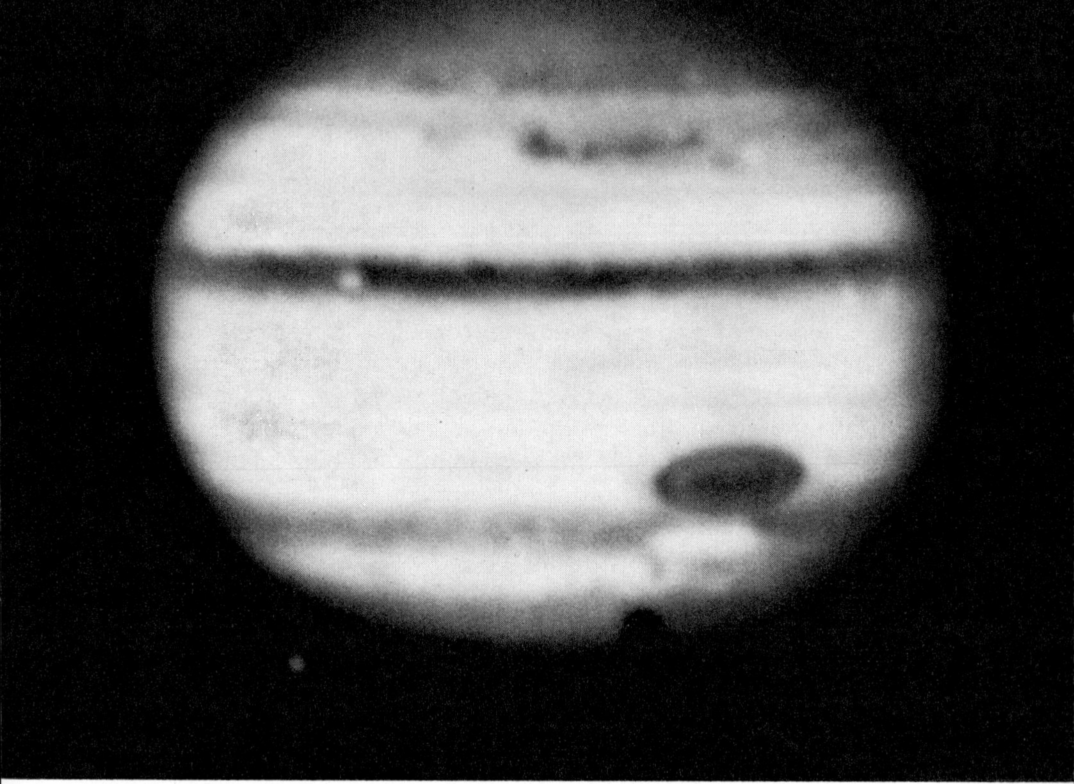

JUPITER, photographed by the 5 m Mount Palomar telescope in blue light to accentuate the dark markings in its atmosphere. Typical cloud bands can be seen, although their appearance changes markedly over quite short periods. The Great Red Spot is prominent in the planet's southern hemisphere. Below and left of the Red Spot, right on the limb, can be seen the round black shadow of the moon Ganymede, itself faintly visible just clear of the planet's lower-left quadrant. Sunlight is coming from the left and the left limb of the planet is sharply defined. Slight gibbous phase is detectable at the right limb.
(Hale Observatories)

The orbits of the Galilean moons lie in the plane of Jupiter's equator and as seen from Earth they appear to move backwards and forward like beads on a string. Once in each circuit each moon passes directly in front of the planet, and then out of sight behind it (although Callisto, the outer moon, can sometimes pass clear of the disc). In a large telescope they can be followed during their transits across the disc, but in the kind of instrument we are concerned with, they disappear. Jupiter, of course, has a long shadow pointing directly away from the Sun and, depending on the angles, a satellite may suddenly disappear from view as it plunges into the shadow, or appear in the same manner as it emerges. This often happens quite a distance from the disc of the planet.

All four would be quite clearly visible to the unaided eye if they were not overwhelmed by the glare of the planet. It therefore takes very little optical aid to reveal them, although Io, being closest to the planet, may require both good conditions and wide separation from Jupiter to be seen in binoculars. In fact, Galileo's discovery was not the instantaneous thing it is widely believed to be: it took him some time and many observations to work out what was involved and Io caused him a great deal of trouble.

Maori Legend

There is a persistent legend among the Maoris and many primitive peoples that Jupiter ('Kopu-nui', and apparently known in some areas as 'Parearau', although this latter name was more usually applied to Saturn) was accompanied by a number of small attendants. Astronomers and physiologists alike would deny that the unaided human eye could detect the satellites because of their small angular separation from the planet and its glare, but perhaps the legends should not be too lightly dismissed.

The dangers of speculation, however, are illustrated by the name 'Parearau' for Saturn. Elsdon Best quotes Stowell as saying the name means 'surrounded by a ring or band' and is 'descriptive of Saturn'. To suggest that any combination of keen sight and good conditions could enable the human eye to distinguish the rings of Saturn, however, is to pass from the possible to the purely fanciful. If it were even remotely possible, it would not have taken European telescopes so many years to solve the problem of Saturn's strange appearance.

The Dynamic Planet

For observers with a telescope, Jupiter itself is accessible and certainly the most interesting and striking of the planets for direct observation. The deep, dense atmosphere of poisonous gases is in constant turmoil, distinctive bands of dark and light clouds being clearly visible. These are relatively stable, but fine detail along their edges and light and dark spots which periodically appear change from night to night, and often during the course of a single observing session. In addition, the rapid rotation of the planet is constantly bringing new parts of the surface under observation.

Most famous of the markings is the Great Red Spot in the planet's southern hemisphere. It has been under observation for more than 120 years and some observations of it go back almost 300 years. It is huge by Earth standards (40,000 km long, half as wide) but is not as big relative to Jupiter as this suggests. For 80 years or more prior to the 1970s the spot showed so little colour that the name seemed a misnomer and astronomers wondered if the older observers had been more enthusiastic than accurate, but about 1971 the colour returned and since then the spot, and much of the planet, have been as bright a pinkish-red as anyone could wish. By 1975, the colour was fading again.

To understand Jupiter and its markings, especially the Great Red Spot, it is necessary to know something of the nature of this unique body. For many years the turbulence of its atmosphere, and the energy levels this revealed, led astronomers to suspect that Jupiter was much warmer than the Sun's heat alone could account for. When, in December 1973, Pioneer 10 became the first man-made object to approach Jupiter, this was confirmed. At low energy (infrared) levels Jupiter radiates as much energy to space from its dark side as from the illuminated face, and in total this is about twice the energy received from the Sun. Jupiter thus radiates as a feeble star, rather than as a planet. The Great Red Spot is almost certainly cloud formations above a giant vortex in the atmosphere, this being the principal mechanism in the southern hemisphere by which heat reaching the base of the atmosphere by conduction from the interior of the planet is transported to the top of the atmosphere by the actual movement of warmed gas.

How true it is to say that Jupiter is a 'failed star' is perhaps debatable, but it is probable that when the Sun and planets condensed from a primeval gas cloud, Jupiter only just missed acquiring enough matter to make it a second star. Later in this book we will encounter the great gas and dust clouds from which stars condense, and learn that usually two or more stars do form close to each other.

Jupiter has a very powerful magnetic field. As the inner big moon, Io, cuts across the lines of force in this field it generates powerful, highly directional radio waves which reach Earth when the angle is right, twice in each orbital revolution of the moon. This radio energy, still not fully understood even after the Pioneer 10 and 11 fly-pasts, has been intensively studied at Otago University.

Jupiter, and indeed all the planets beyond Earth, is best observed near what is termed 'opposition', when Sun, Earth and planet are in a straight line and we see the planet rise as the Sun sets. It is in the sky all night and sets at dawn. Jupiter comes to opposition every thirteen months, so there is an opposition in most years (but not in the year following a December opposition). When observing a particular feature and seeking to follow it over a number of nights, remember the ten-hour rotation period. Thus, if you are looking at the Great Red Spot at 8 p.m., you can see it again ten hours later (6 a.m.) in the same position. Ten hours after that, however, is 4 p.m., and although the spot will be facing Earth, it is not visible in broad daylight. To see it again, you will have to get up at 2 a.m. the next morning, and so on.

Chapter Six

THE LESSER GIANTS AND BEYOND

The Beautiful Planet

If Jupiter is the dynamic planet, Saturn is the beautiful one. In even a large telescope, the ball of the planet is unremarkable—a few faint bands, an occasional light spot, an even rarer dark one—but it is not for the planet itself that we view Saturn. The prime attraction is the ring system.

To this first of the worlds beyond Jupiter belong a number of distinctions. It is the most distant planet observable with the unaided eye and thus, until 1781, represented the outer boundary of the solar system; and it is the first and largest of the three giant planets which form the third of the four groups into which we divided the planets of the solar system.

Saturn takes almost thirty years to complete its orbit, 1,427,010,000 km from the Sun. It is 119,300 km in diameter and more than ninety-five times the mass of Earth. It rotates on its axis in 10 hr 14 min. The ring system is 273,000 km across. Compare this with the Earth-Moon distance of 385,000 km.

Saturn, and the other two members of its group, Uranus and Neptune, are built on a model that sharply distinguishes them from the Earth-like planets. Theory suggests they have a rocky core, bigger than Earth but not excessively so. Because of their size and distance from the Sun, however, they have been able to retain their hydrogen and other light gases. These overlie the core in thick layers of frozen material, covered in turn by liquids and dense atmospheres of poisonous gases. Because so much of their material is light, whether frozen or gas, these planets have very low densities—indeed, Saturn's is less than that of water. This leads many writers to point out that the planet would float if placed in a sufficiently large ocean. They never say where such an ocean might be found.

Bright, Divided Ring

Saturn is surrounded by a system of rings, divided into three zones (although only two are easily observed) and lying in the plane of the planet's equator. The rings are composed of small particles orbiting the planet like a host of tiny moons, and not as close together as the solid appearance of the rings would suggest. The rings are very bright, so bright that the particles must either be

SATURN, with its ring system as wide open as we ever see it. The two main zones of the ring, separated by Cassini's Division, are clearly seen. The third zone, also known as the Crepe Ring, is the innermost zone of the three. It is extremely faint and can be seen only in good conditions in an instrument of moderate aperture. All detail is much fainter on Saturn than Jupiter and apart from a light equatorial zone and a slight darkening toward the poles, little is usually visible in a small instrument. When Earth is much closer to the plane of the ring system than it was when this photograph was taken by the 2.5 m telescope at Mount Wilson, the ring appears more nearly closed, and both poles of the planet are visible. *(Hale Observatories)*

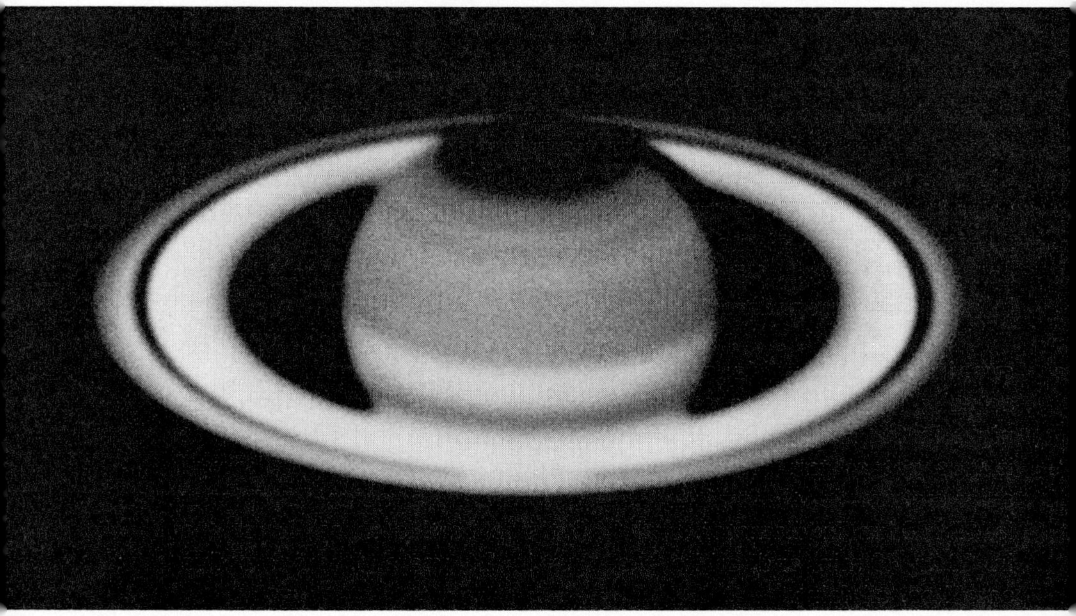

ice, or ice or frost-coated (possibly, though not necessarily, water ice). The two brightest rings, or zones of the ring system, are separated by a dark, narrow gap called Cassini's Division.

The angle at which we view the ring system is constantly changing, and this is important to users of small instruments. When Earth passes through the plane of the rings, they are edge-on to us and disappear in all but the biggest telescopes. When the angle between Earth and the ring plane is at its greatest, they are as wide open as we ever see them. At such times the planet is a distinctly elliptical object even in binoculars. and a small telescope will show the rings.

Saturn occasionally passes in front of a relatively bright star (although much less frequently than might be thought), and if part of the ring system also happens to pass in front of the star it is a fascinating sight. The star can be plainly seen to brighten and fade as its light is obscured by parts of the ring.

Saturn has ten known satellites, of which several of the brightest are visible in small instruments, but, being much further away than the moons of Jupiter, are much fainter. They are not too difficult to identify in a 75 mm telescope.

The 'Modern' Planets

Saturn is the last of the planets known to the ancients. Beyond it lie two giants and a dwarf, all modern in the sense that they were discovered long after astronomy had become an established science. None is of interest to observers with small instruments, although the two giants, Uranus and Neptune, are easily detected as star-like points of light in binoculars when near opposition and in good seeing conditions.

Uranus, in fact, is just bright enough to be visible to the unaided eye, in perfect conditions and if the observer is experienced. It was actually recorded as a star almost a score of times before that night in 1781 when Sir William Herschel saw it, and realised that it was not a star. This discovery doubled the size of the known solar system, for the orbit is 2,869,000,000 km in radius. The planet is 47,100 km in diameter, and 14.6 times Earth's mass. It rotates in 10 hr 49 min and orbits the Sun in 84 years. There are five satellites, all faint.

A Triumph of Science

Neptune was discovered by J. Galle of Berlin Observatory in 1846. Behind this discovery—one of the earliest and one of the greatest triumphs of the 'scientific method' in any branch of science—lay a triumph and a tragedy. The story began when Uranus was found to depart from its calculated orbit. The most likely explanation was that it was being attracted ('perturbed') by a more distant planet. From the available data, two mathematicians, J. C. Adams in England and U. J. J. Leverrier in France, set out independently to calculate the position of the unknown planet. Both were successful. Leverrier communicated his results to the Académie des Sciences, Paris, and had them immediately accepted and made available to observatories. Galle found the planet three weeks later.

In England, Adams communicated his results to the Astronomer Royal, Sir George Airy, who at first didn't take the work of a young man in so difficult a field seriously and did nothing. His delay was fatal. The search had barely begun in England when Galle's discovery was announced. It is worth recording that neither Adams nor Leverrier took part in the bitter recriminations that followed as to who deserved the credit. Today, both are remembered with equal honour.

Neptune is 4,496,700,000 km from the Sun, 48,400 km in diameter and 17.2 times Earth's mass. It rotates in 15 hr 48 min and revolves about the Sun in 165 years. It will not be back in the position in which it was discovered until the year 2011. Neptune has two satellites.

The Gate-crasher?

Pluto is a dwarf, Earth-type planet orbiting 5,796,000,000 km from the Sun. In even the largest telescopes it barely shows a disc, so diameter and mass are uncertain. They are thought to be 5900 km and 0.08 of Earth respectively. Pluto is a little larger than the giant satellites of Jupiter and Saturn.

Pluto may be a gate-crasher. One of the difficult things to explain about the solar system is why there is an Earth-type planet in such a position. Another is how it happens that Pluto *isn't always* the most distant planet from the Sun: for a small segment of its orbit it is closer to the Sun than Neptune. The inclination of the orbits, however, ensures that the two planets will never collide.

Only one theory accounts for all the facts: that Pluto was once a third satellite of Neptune that escaped from that system and fell into an orbit around the Sun. Pluto is moving toward perihelion (the point in its orbit at which a planet is closest to the Sun) now. About 1979 it will cease to be the most distant planet, ceding that honour to Neptune for the next 20 years. Perihelion is in 1989.

Beyond Pluto

It is fascinating to speculate on whether a tenth planet slowly circles the Sun in a huge orbit far beyond Pluto. The possibility has been suggested many times and based on more than one kind of evidence, but no definite indication of such a planet has yet been found.

One basis for the prediction is Bode's Law, which isn't really a law, but a numerical relationship between the planets' distances from the Sun. It was thought to be good evidence for the probable existence of a planet between Mars and Jupiter and was the reason for the search which culminated in the discovery of the first of the asteroids. It also predicted the positions of Uranus and Neptune with surprising accuracy prior to their discovery. Bode's Law can, of course, be extended to any number of hypothetical orbits and planets do not necessarily exist in them.

More material evidence, however, is an apparent tendency for a number of comet orbits to have their greatest distance from the Sun close to the distance at which Bode's Law would place a tenth planet. If such a planet did exist, it would neatly account for the comet orbits which extend no further, but this evidence is suspect; the effect may not be real. Comets are very numerous and all have not yet been discovered. As further discoveries are made, the apparent clustering may disappear as more and more comets are found to reach their greatest distance from the Sun (aphelion) on each side of the apparent cluster.

A group of short-period comets is associated with Jupiter in this way, but increasing discoveries have shown that groups which appeared to be associated with Saturn, Uranus and Neptune are not real; they have merged into an even distribution.

Other evidence sometimes quoted in support of a tenth planet is now definitely known to be fallacious. It is based on supposedly unexplained disturbances to Uranus, but these are now known to have been the product of inaccurate positions used for early calculations of the orbit. The only influence from further out which disturbs the orbital motion of Uranus is that caused by the planet discovered because of it: Neptune. The remaining unexplained disturbances which led to the search for Pluto are now thought to have been the result of error, making the finding of Pluto close to the predicted position one of the most amazing coincidences in scientific discovery.

If there is a tenth planet, no definite evidence for its existence has been found and even if there were a giant comparable with Uranus and Neptune, its discovery at such a distance would pose major problems for Earth-bound equipment. Inter-planetary probes escaping from the solar system after close approaches to the outer planets might conceivably find it, but it is doubtful that the possibility is taken seriously enough for probes to be programmed to conduct such a search.

Chapter Seven

THE COMETS AND METEORS

Two other kinds of object complete the Sun's family—comets and meteors. Neither have any great mass or substance, but both can, on occasions, be extremely spectacular. Comets and meteors are related to some extent, and have several characteristics in common.

Both exist in very large numbers and have orbits which are not confined to the plane in which we find the planets. The orbits are orientated at every conceivable angle around the Sun and, particularly in the case of comets, are usually long, narrow ellipses that keep the object far from the Sun for most of its orbital period. Both are probably debris left over from the formation of the planets. The main difference is that meteors are entirely stony or metallic, while comets are believed to be 'dusty snowballs'.

The Comets

Comets are at most a few kilometres in diameter, frozen solid while far distant from the Sun, but vapourised to gas and dust by solar heat during the once-in-an-orbit close approach.

Many comets recede to great distances, some to many times the radius of the orbit of Pluto, but a few are in short-period orbits, always relatively close to the Sun. Because their mass is small, they are at the mercy of the gravitational pull of the planets and their orbits are often changed considerably over long periods. When a comet has made a number of close approaches to the Sun or has been too long in a short-period orbit, its volatile material is exhausted and its rocky core, if it has one, cannot be distinguished from an asteroid.

How spectacular a comet appears to us depends upon a number of factors: how big it is, how much volatile material it contains, how closely it approaches the Sun, how close it is to Earth and the angle at which we view it. The distances, particularly that from Earth to the comet, and the angle can be different for successive returns of the same comet so that it may be spectacular on one occasion and far less so on another.

About twelve new comets are discovered each year which, together with known comets making predicted returns, means that there are usually several in the sky at any given time, but the majority are very faint, able to be photographed with big telescopes but not accessible to small instruments or the unaided eye. A big, bright comet is comparatively rare.

The most famous comet of all, Halley's Comet, last made a close approach to the Sun and was seen from Earth in 1910. It then began a retreat from the Sun along its narrow orbit that carried it beyond the orbit of Neptune some time prior to 1930. By this time it was travelling very slowly; it was 1948, and many millions of kilometres beyond Neptune, before it turned back toward the Sun. Ever since then it has been gradually gathering speed, but it was 1967 before it was again inside the orbit of Neptune. By 1975 it was two-thirds of the way from Neptune to the orbit of Uranus and in 1980 will be about a quarter of the way from Uranus to the orbit of Saturn. It is obvious that it moves very slowly at these distances from the Sun.

After 1980, the steady increase in speed will be far more noticeable and in only another four years the comet will have almost reached Jupiter. By that time it will probably have been photographed from Earth, and some time in 1985 it should become visible in small telescopes. Closest approach to the Sun is in 1986, after which it will begin again its long journey to the cold and dark of the outer reaches of the solar system beyond the orbit of Neptune.

This great comet is unique in history. It was in the skies of Earth to mark events as diverse as the death of the Roman Emperor Macrinus, the death of Attila the Hun and the Norman Conquest of England, and it was the vehicle of a great advance in astronomical science.

Edmund Halley, armed with Isaac Newton's newly discovered

law of gravitation, was able to look at the chaotic vista of history's comets, which many men had studied, and see what no man previously had seen: that the great comets of 1531, 1607, and 1682 were moving in orbits so similar that they might well be a single body with an orbital period of 75 to 76 years.

On this evidence, Halley made in 1705 the prediction that the comet would return again in 1758. Halley died in 1742. As 1758 approached an intense interest gripped the world, an interest which an astronomical event has probably attracted on only two other occasions: the discovery of the planet Uranus by William Herschel in 1781 and the Moon landing in 1969. In the second half of 1758 almost every person on Earth with access to a telescope was searching for the comet, and on Christmas night it was found by an amateur astronomer, George Palitzsch. Edmund Halley's fame was assured, and comets were removed forever from the realm of the supernatural and placed where they belong, among the Sun's family, subservient to the same laws of gravity and physics that control the planets.

New Zealanders have a better reason than the comet to remember Halley, for it was as a result of some of his other work that Captain Cook was originally sent into the Pacific (see Appendix Three).

As a comet approaches the Sun, the ice begins to turn to gas and releases also any dust which it contains. This forms a cloud called the 'coma' around the solid head. The material of the coma then begins to stream away from the comet under the influence of the Sun's radiation. The atoms and molecules of gas are driven directly away from the Sun, but particles of greater mass are affected to some extent by its gravity. Thus, the gas tail points always away from the Sun, but the dust tail may be quite perceptibly curved. Some comets show both types of tail, some only one.

Meteors and Meteor Explosions

It is unlikely that every meteor was once the core of a comet, but some undoubtedly were. These wanderers are extremely common among the planets, but the majority are small. The craters of the Moon are believed to be meteor impact craters, or, more correctly,

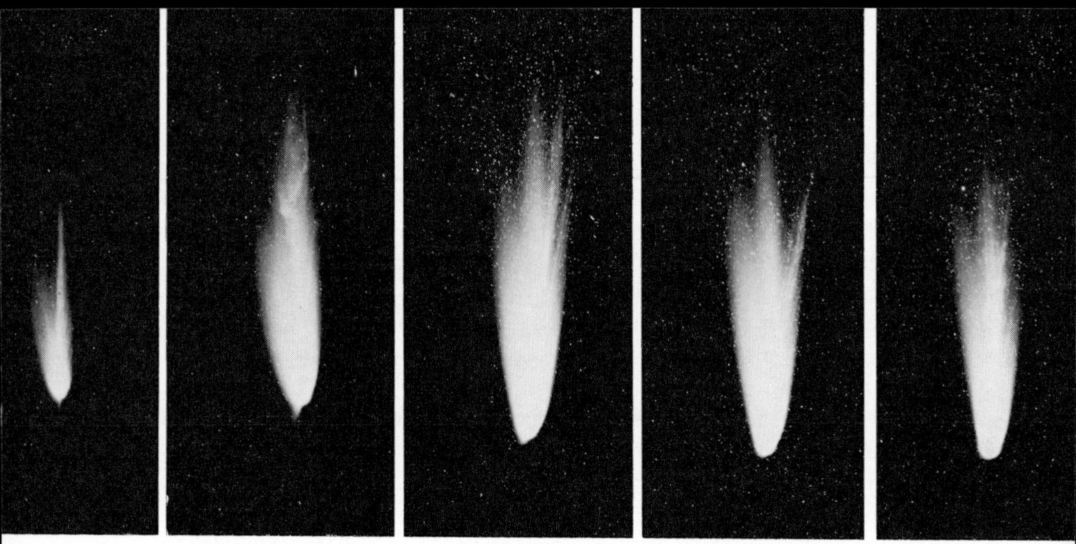

April 26 April 27 April 29 April 30 May 1

COMETS AREND-ROLAND (above) and MRKOS (below) photographed in 1957 by the 1.2 m Schmidt camera on Mount Palomar. Neither is ever as spectacular unaided or in a small instrument as these photographs suggest, but they show a number of significant comet characteristics. In each case a gas tail can be clearly seen, driven directly away from the Sun by radiation pressure. Other tails, several in the case of Arend-Roland, are formed of dust particles moving partly under the influence of radiation pressure and partly under the Sun's gravity. The dust tails are curved, although this does not always show because of the angle at which the comet is viewed. It can be clearly seen that the gas tail contains internal structures and that these change quite rapidly as you'll note by the dates. The spike that appears to project forward from the head of Arend-Roland in the views of 26 and 27 April is actually an edge-on view of material spread around the head in the plane of the orbit. *(Hale Observatories)*

August 22 August 24 August 26 August 27

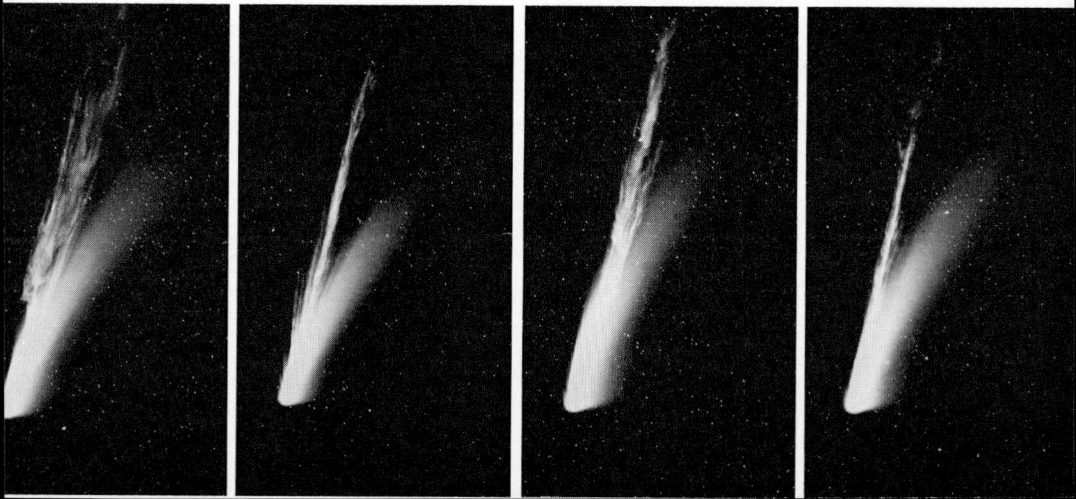

explosion pits caused by meteors. One of the old objections to the meteor theory of the origin of lunar craters was that, because the craters were circular, every meteor would appear to have arrived on a course perpendicular to the surface, and this was most unlikely. At that time it was thought that a meteor would cause only a simple impact pit.

It was A. W. Bickerton, Professor of Chemistry at Canterbury University in the latter years of the nineteenth century, who first directed attention to the physics of collisions between celestial bodies, although in another context. His friend, A. C. Gifford, then science master at Wellington College, did the calculations that showed that, on impact, the 'kinetic energy' (energy of motion) of the meteor is transformed into heat, vapourising the meteor. The hot gases then explode violently outward, causing an explosion pit with all the characteristics commonly found in lunar craters. This was not the first time this had been suggested, but a paper published earlier had not attracted much attention. It was the work of Bickerton and Gifford that placed the meteor theory on a firm basis and, although today still not fully proved, it remains the most likely explanation for the craters of the Moon.

When a meteor collides with a body protected by an atmosphere, as is Earth, a different sequence of events takes place. Friction with the air rapidly raises the outer surface of the meteor to incandescence and it begins to stream away, often leaving a trail clearly visible if the event takes place in darkness. The majority of meteors, being small, are quickly burnt up, but a few of the larger ones survive long enough to reach the surface of the Earth, like the one that landed at Mokoia in South Taranaki in 1908, or the much larger one which several thousand years ago blasted the huge Barringer Crater in the Arizona Desert. The Canterbury Museum holds a fragment of this metallic meteor that is the third largest ever found and the largest ever to leave the United States.

Meteor Showers

Some meteors derived from comets are members of large amounts of material spread out along a single orbit. When Earth passes through one of these meteor streams, very large numbers may

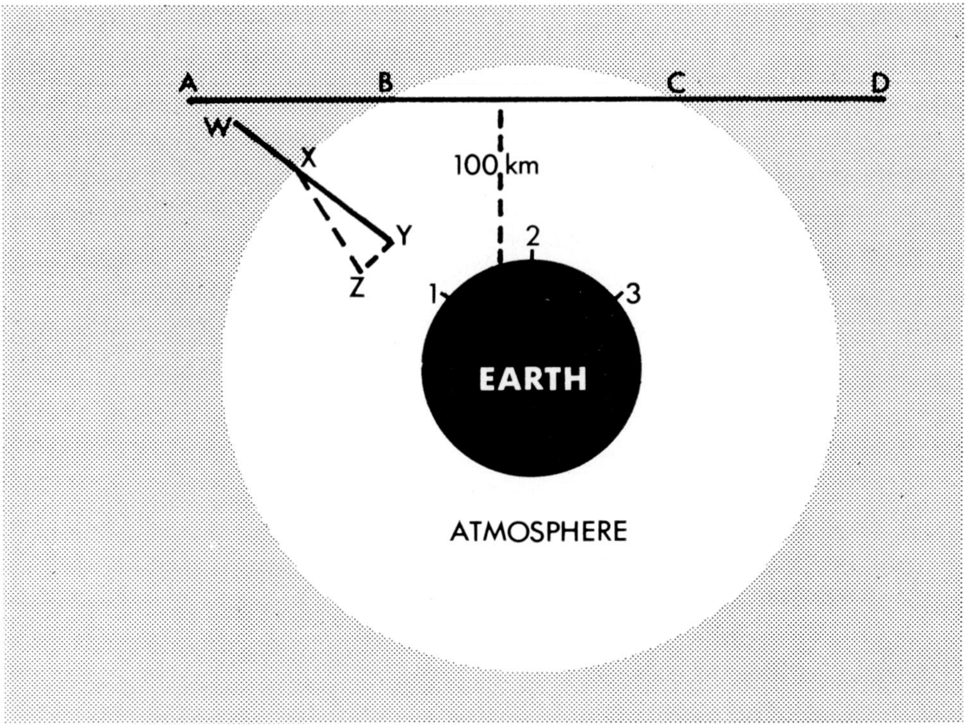

METEOR PHENOMENA (not to scale). Meteors can appear in a number of guises, some of them unexpected. It is important to remember that, because they are almost always about 100 km up, perspective ensures that what we see is often only apparent, not real. Other effects follow naturally, provided the circumstances are understood. One of the least understood of meteor phenomena is that a meteor can appear as a rising light. (Many flying saucer reports are endorsed with the words: 'It was a rising light, so it couldn't have been a meteor.') Consider the meteor track A-D, which shows a meteor passing right through Earth's atmosphere and emerging into space again, a frequent occurrence. It is visible between points B and C. Three observers viewing it from different positions will see quite different things. Observer One will see it appear high in the sky and fall to the horizon, where it passes out of sight. He will have little doubt that it is a meteor. Observer Two will see it pass horizontally through the sky above him, and he will probably realise what it is. Observer Three, however, will see it appear at the horizon and climb high into the sky, then disappear. Unless he already knows something about meteors, he is likely to seek a far more spectacular explanation and may well end up reporting a flying saucer.

The second meteor track illustrates the difference between the real track and the apparent one. The meteor traverses the track W-Y, becoming visible at X and burning out at Y. Because Y is likely to be 75 km high, the observer sees it apparently move from Z to Y, a short apparent track compared with the real track X-Y. The apparent speed along Z-Y will be much slower than the real speed along X-Y, and may encourage the observer to believe that only a powered craft could move so slowly, if he does not realise that perspective is playing tricks with a quite simple meteor sighting. Several meteors travelling in the same track may enter the atmosphere in quick succession, possibly so close together that they are simultaneously visible. A train of three to five bright meteors standing majestically across the sky is a magnificent sight—and a common source of flying saucer reports of the 'long, cigar-shaped body with a row of illuminated ports' variety.

enter the atmosphere. These meteors are fragile clumps of dust, quite different from the stony or metallic meteors. During these 'meteor showers', many trails may be visible simultaneously for hours at a time, sometimes for several nights in a row. The trails are in fact parallel, but perspective causes them to appear to radiate from a point, the 'radiant', and the shower is usually named for the constellation in which the radiant appears (see Appendix Two).

INTERLUDE

The Forgotten Chemist

Professor A. W. Bickerton was keenly interested in astronomy and his almost-forgotten role in lunar crater theory was his greatest contribution to it—but he died believing he had made a far greater one.

In his day the major problem of astronomy was the phenomenon of the 'nova', the 'new star' that blazed out for a few months, then faded rapidly into obscurity again. The astronomy of the time could cope with the apparent birth of a new star, but not with its rapid disappearance.

Professor Bickerton suggested that perhaps a large number of stars, even a majority, were dark, or at least too faint to be detected. If this were so, collisions between them might be relatively frequent and the nova might be the friction-heated debris torn from two such stars in a grazing collision.

This was an almost perfect theory; it took observed evidence, made only one assumption and drew a logical conclusion. In fact, the assumption that a significant part of the mass of the galaxy is not in the form of bright stars was entirely correct, although we know now that it consists of dust and gas, not dark stars.

His suggestion was at first hailed as brilliant, but shortly afterwards it began to be increasingly apparent that he was wrong. It was not until the 1960s, however, that a better explanation was evolved, and throughout his life Professor Bickerton refused to disown his brainchild. He became increasingly abusive of his opponents and when he died he was remembered mainly as a crank astronomer.

It has been almost entirely forgotten that he was not an astronomer, but a teacher of chemistry. Should we not judge a teacher by the accomplishments of his students and give Professor Bickerton his share of the credit for that Nobel Prize in Chemistry which in 1908 was conferred upon one of his students who was not principally a chemist, but a physicist? The student whose name was Ernest Rutherford.

Chapter Eight

THE NUCLEAR FURNACE

At the heart of the solar system is the Sun, under whose gravitational control all the lesser bodies of the system move and from which they can never escape. The total mass of all the bodies that orbit the Sun—the planets, asteroids, comets and meteors—is only a tiny fraction of the Sun's bulk. Compared with it, they are totally insignificant.

The Sun is a thermo-nuclear reactor more than 1.3 million kilometres in diameter, pouring its radiation into space around it, the sole source of light and heat for all its family. About 5000 million years ago it began to condense out of a cloud of primeval gas and dust; millions of years later its internal temperature grew high enough for nuclear reactions to begin in the hydrogen at its core. It became very hot, then settled down to that steady state in which we see it today, and which it will maintain for many millions of years.

The Death of the Sun

The Sun will not last for ever. Between 5000 million and 10,000 million years from now, it will begin to exhaust the hydrogen that forms almost all of its present mass. It will then consist mostly of helium, the 'ash' from its nuclear furnace, which is already building up at the core. In the next stage, the core will collapse and the helium begin to 'burn' in a new sequence of thermo-nuclear activity. The outer layers will expand, probably consuming the Earth, and the Sun will become a 'red giant'. This phase will be quite short before a final collapse occurs and the Sun becomes so small that even from as close as the Earth now is, it will be only a point image in the sky, blindingly bright, but showing no disc. It will then be a 'white dwarf' star, its collapsed material so dense that a few cubic centimetres will weigh many tonnes.

For many stars, particularly those more massive than the Sun, the end is quite different.

All this lies so many millions of years in the future that the time involved is quite meaningless, yet astronomers feel reasonably sure that this is what will happen because the evidence is all around us. The Sun is only one of countless millions of dwarf stars in our galaxy and others. Many are younger than the Sun; many are older. In them all, the history of any one of them may be read. Some stars are smaller than the Sun, some are bigger, but of the vast majority of stars in the universe, the Sun seems to be typical. Yet in one respect it is unique.

Of all the stars in the universe, the Sun is the only one close enough to show us its surface, the only star of which we can see more than a point image. Because it is so close, we can find out more about the Sun than we can ever hope to about any other star—and everything we learn about the Sun is applicable in some way to most other stars.

Be Very Careful

It is possible to discover something of the Sun's characteristics and behaviour with small instruments, but first a warning that must be heeded.

The Sun is the most dangerous object in a volume of space stretching at least two light-years in every direction. We are only a few million kilometres from it, most inadequately protected by a tenuous atmosphere. The surface of the Sun is at a temperature of 5500 degrees Celsius. It is too bright to view directly without risk of injury. *Instant blindness will result from using a telescope or binoculars without the precautions described here.*

Astronomers seldom draw attention to interesting solar phenomena such as giant sunspots because, despite the most dire warnings, many people fail to take precautions and suffer eye damage. Partial and total eclipses draw attention to themselves, and are always followed by numerous cases of eye damage. One of the factors commonly misunderstood is that when most of the Sun has been obscured by the Moon, it may be possible to stare at it without discomfort, for the blinding volume of light has been reduced. The intensity, however, has not, and it is this which does the damage.

Stringent Precautions

It is necessary to take stringent precautions to ensure that an instrument is *never* pointed at the Sun, even by accident. If using binoculars to search for Venus in daylight, for instance, always ensure that the Sun is hidden behind a roof or tree, thus removing any possibility of mishap (in any case, it makes it easier to find the planet).

Astronomers are especially worried because many small telescopes, particularly Japanese ones, are sold with 'solar filters' that are almost certainly dangerous, especially those used at the eyepiece. To be effective, a filter must cut out not only visible light, but also the invisible ultraviolet and infrared that are actually more damaging. A filter that will do all this is a rare and expensive piece of glass, and the cost of the telescope is proof enough that the filter is not in this class. Any filter, however, absorbs the radiation it does not pass. This means that a solar filter is heating up the whole time it is in use, and eventually will crack. If damaged or in any way defective, it may shatter at quite low temperatures. No filter is completely safe.

Alternatives

There are alternatives. One is an expensive piece of equipment called a 'Herschel wedge', placed between eyepiece and telescope. A prism reflects out of the system 99 per cent of the light falling on it and allows only 1 per cent to reach the eyepiece, but even this 1 per cent is first passed through a dense filter. When using such a wedge, the whole system heats up and can only safely be used for a few minutes at a time.

The second method is projection—using the telescope to project an image of the Sun on a light-coloured surface. A suitable screen is placed behind the eyepiece and by varying the screen-eyepiece distance and the focus of the eyepiece, an image of the Sun can be focussed on the screen. The most satisfactory method is to fit the screen across one end of an open-ended box, the other end of which fits snugly around the telescope tube. A suitably thin material is used for the screen, which is viewed from the back.

This second system is safe provided two things are remembered: the whole telescope will heat up with the solar radiation pouring through it (although this is no problem with a small instrument), and secondly, if the eyepiece is not fully enclosed by a box of the type sugested, someone who does not appreciate the danger, especially a child, may sneak a quick look with disastrous results. It is a further advantage of projection that quite a large group can view the image at one time, making it ideal for showing sunspots to the public or school parties.

Observing Sunspots

Sunspots wax and wane over what appears to be an 11-year cycle. (We won't go into details, but this is actually half a 22-year cycle determined by magnetic, but no visible, effects.) Few spots can be found at minima, but progress toward maxima is marked by a steady increase in both numbers and average size.

The biggest sunspots are many times the diameter of Earth and may be clearly visible to the unaided eye if a sufficiently dense filter is used (the densest grade of filter used for the window of an electric welder's face mask is suitable). Spots mostly occur in pairs, although one component may be far less prominent than the other. They are the open ends of 'tubes' caused by magnetic fields in the outer material of the Sun and appear dark because they are cooler than the incandescent gases which surround them, but this is only a contrast effect. The spots are about 3000 degrees Celsius, 2500 degrees cooler than the surrounding material.

Solar eclipses (see also the section on the Moon) are interesting, although total eclipses are not often seen and there is not another in New Zealand this century. The Sun is the only body bright enough to make photography simple, but all the problems caused by heat remain. Sunspots and all phases of an eclipse other than totality may be easily and safely photographed on the projection screen. The total phase may be photographed directly, though a focal length normally only found in a telescope is required to yield an image of useful size. Using a camera behind an unfiltered eyepiece is to risk damage to film and camera, particularly the shutter.

An amateur observer can no longer do useful work by keeping sunspot counts, but the Sun remains an interesting object for study—provided precautions are taken.

Chapter Nine

THE STARS: THE MILKY WAY AND BEYOND

The time has come to make the jump across the vast gulfs that separate the Sun from the other stars that stretch for light-years in all directions around us. All the stars that we can see, distant though many of them are, are members of a single system called the Milky Way Galaxy. Beyond its boundaries lie other galaxies, two so close that they are satellites of it, a few more forming a local cluster, and beyond them galaxies seemingly without number, out as far as telescopes can penetrate. Before we look at individual objects in the Milky Way Galaxy, we will look at it as a whole, and try to form some idea of its shape and structure. Later, we will compare it with more distant galaxies.

If we could travel far outside the Milky Way Galaxy and circle around it, we should find that its appearance changed as our viewpoint shifted. From one direction we would see a long spindle, with a glowing, bulbous centre, the long axis of the spindle marked by dark filaments and streamers. Examined in a telescope, the glowing areas would be found, in some cases, to be star clouds, in others, clouds of glowing gas and dust. The dark areas are also gas and dust, nebulae which contain no stars to make them shine and which obscure the bright areas beyond them.

Moving our viewpoint through 90 degrees, however, would reveal a totally different appearance. Then the galaxy would be a wheel seen full-on, the hub a glowing star cloud wreathed in gas and dust, the spokes great spiral arms radiating from the hub (nucleus) and marked out by dense groupings of stars (many of them young, hot and blue) and the glowing gas and dust clouds in which still more stars are condensing. So striking is this great wheel that it would be easy to conclude that it is all there is to the galaxy. Careful examination would show us two things, however: one, the space between the arms is not empty and, two, the wheel is only the central plane of a spherical structure.

Between the spokes of the wheel are many stars, although not so numerous as in the spokes (which from now on we will call spiral arms). These stars are mostly smaller, older and more red in colour than the young giants of the arms and the space around them is almost entirely free of the gas and dust which characterises the arms. Few new stars are forming in such regions. This region fills a spherical volume of space, the arms lying in a disc through the equatorial plane. Also spaced through this spherical volume are some of the most spectacular objects in the galaxy, huge clusters of stars numbering from hundreds of thousands to millions: the Globular Clusters.

With our unaided eyes, we can trace part of the structure of the Milky Way Galaxy in the sky, see typical examples of the types of object it contains and, far beyond its borders, see other galaxies, many whose structure is basically different from our own. With binoculars or a small telescope, these views are enhanced, and more objects may be seen.

SPIRAL GALAXIES NGC 4594 (overleaf left) and NGC 5194 (overleaf right), each essentially similar to our Milky Way Galaxy, photographed by the 5 m Mount Palomar telescope. The central regions of our galaxy are not as bright as the nucleus of NGC 4594. This galaxy is in Virgo and we see it edge-on to the disc, so its spiral arms are not visible. NGC 5194 is in Canes Venatici. We see it full on, and the spiral arms stand out prominently. The rotation of the galaxies has concentrated gas and dust in lanes in the equatorial disc and the formation of stars has proceeded rapidly in the concentrations. The arms, whether seen full or edge on, are clearly marked by large numbers of predominantly young and hot stars and by the gas clouds, both bright and dark, in which stars are still condensing. Between the arms are fewer stars, mostly older and more red, and these extend also through the spherical volume of the galaxy, although no trace of this thinly populated region can be detected in either photograph. In the spherical volume ('halo') are found also the globular clusters. These two galaxies can be regarded as looking rather like the Milky Way Galaxy would look if we could travel far outside it and see it from positions in the plane of the central disc, and at right angles to it. NGC 5194 even has a Magellanic Cloud-like companion galaxy, although no filament of stars and gas joins the Clouds to the Milky Way Galaxy.

(Hale Observatories)

The Constellations

Before going further, however, we will consider the system of constellations and star designations, or, approached piecemeal, the system will appear confusing. Basically the system is simple, but there are some exceptions and a few inconsistencies to complicate it slightly.

The whole sky is divided into eighty-eight constellations. All the northern and some of the southern have borne since antiquity the names of mythical figures, beasts or objects—Orion the Hunter, Taurus the Bull, Libra the Scales. Most of the southern constellations have been named in comparatively modern times and some have a distinctly contrived sound—Telescopium the Telescope, Sculptor, the Tools of the Sculptor, and so on. In some cases the name is one word, in others two. Among the double names there are many pairs where the first word is the same in each and the second part is a word such as 'Major', 'Minor' or 'Australis', in each case signifying two constellations in different parts of the sky. Some names are deceptively similar: Canes and Canis, Sagitta and Sagittarius.

The list of eighty-eight constellations is occasionally increased to eighty-nine by references to Argo, a big, old constellation now officially divided into four (Pyxis, Puppis, Vela and Carina). The boundaries of all constellations lie on lines of Right Ascension and Declination, the 'longitude' and 'latitude' of the sky. Many constellations, however, are most irregular in outline and a few are unusual. Thus, the Southern Cross is a small constellation surrounded by Centaurus, and Serpens is divided into two separate parts by Ophiuchus. The name of a constellation is given in Latin, thus 'Crux Australis', not 'Southern Cross', although the latter is correct for the Cross itself as distinct from the whole constellation.

Individual Designations

Within each constellation individual objects are designated by a system which becomes a little complex as objects become fainter and fainter. The method is to use a Greek or Roman letter, followed by the name of the constellation in its genitive form (for example, in the constellation Centaurus the brightest star is Alpha

Centauri). The letters used are, first, the small letters of the Greek alphabet, then, when those are used up, small Roman letters, and finally Roman capitals. In practice, all bright objects in the richest constellation have been disposed of by the time the Roman capital 'Q' is reached, and Roman capitals after 'Q' are used to designate variable stars (which are so numerous that designations have in some constellations reached double capitals such as 'BZ'). The faintest stars are identified by their number in the official astronomical catalogues, such as the 'New General Catalogue' (NGC), but few of these will concern us.

The letter sequence usually begins with the brightest star in the constellation and proceeds through the fainter stars in order. Thus, the brightest star in Crux Australis is Alpha Crucis, the second brightest Beta Crucis, and so on. There are exceptions, however. A natural sequence, or position in the constellation, was sometimes preferred; the brightest star in Orion (Rigel) is Beta Orionis and the second brightest (Betelgeuse) is Alpha Orionis. Occasionally a bright star given a Greek letter proved to be variable and upset the sequence, and the division of the old constellation Argo into four has left some inconsistencies in its wake.

Typical Inconsistencies

The star originally designated Eta Argus illustrates both points. It is now Eta Carinae and much fainter than when it earned itself a Greek letter when John Herschel classified it. Canopus, the brightest star in Carina (and second brightest in the sky) was originally Alpha Argus—and usually still is, although it is sometimes given as Alpha Carinae.

A number of the brightest stars have proper names, and the better-known ones are commonly used. Thus, the two brightest stars in Orion are frequently referred to as Rigel and Betelgeuse and the northern Twins as Castor and Pollux. The three stars of Orion's Belt, however, although they have names, are usually referred to by their Greek letters, and names such as Rigil Kent for Alpha Centauri are found in lists of star names but never used. These unusual names appear to have been taught in the Services for astro-navigation during World War II, perhaps because they were easier to learn and more distinctive than the Greek letter designations.

There is no need to learn the system, provided its broad basis is understood and the fact that there are inconsistencies and irregularities is remembered. The genitive form of the constellation name is in most cases obviously derived from the usual form, but a few are not as close as might be expected. Lastly, two names have alternative forms: Scorpius is sometimes given as Scorpio, and Camelopardalis sometimes as Camelopardus.

Size, Luminosity and Mass

Stars vary tremendously in size and in brightness. There are diffuse red giants, such as Betelgeuse, Aldebaran and Antares, whose diameter can equal that of the orbit of Mars or Jupiter, but whose average density is so low that we would regard it as a very good vacuum if we could produce it in a laboratory. There are blue giants, such as Rigel, smaller than the red giants, but still big enough to dwarf the Sun. There are red dwarfs, not much bigger than Jupiter, and white dwarfs and neutron stars 10 km or less in diameter. Some red dwarfs are like faint candles guttering in the void. The blue giants can be seen across the gulfs that separate the galaxies; and supergiants are many times more luminous. These size and luminosity differences are in ratios of hundreds of thousands to one, but the masses of most stars lie between one-tenth of the mass of the Sun and ten times its mass. This is a mass ratio of only 100:1. Very few stars are bigger or smaller than this.

Colour Among the Stars

A star's colour is a product of its temperature, or rather, the temperature of the outer layers that we see, for the interiors are immensely hotter. The cooler the star, the more its light will tend toward the red end of the spectrum; the hotter the star, the more its light will approach blue. Thus, the coolest stars of all are red (whether they be very young or very old) and the hottest are blue.

Unless the star is also bright, its colour will not be easily distinguished in a small instrument. It is true that John Herschel called the Kappa Crucis Cluster (adjacent to the Southern Cross) the Jewel Box from the many colours that can be seen among its stars, and certainly in good conditions it can look like a handful of precious stones scattered on black velvet around the deep red ruby of Kappa Crucis, but with less than good conditions, or with an inadequate instrument, it will disappoint.

The bigger the telescope, the better the chance of seeing colour in faint stars, but on the other hand, the bigger the instrument, the better the seeing conditions will have to be to permit it to perform well. Judged solely from the consideration of seeing colour, a reflecting telescope is superior to a refractor.

A refractor does not bring light of different wavelengths (colours) to the same focus, and even in an expensive, well-corrected instrument some residual distorting colour effects remain. Colours are *reflected* equally so this problem does not arise when the objective is a mirror. But colour distortion may still appear in the eyepiece, especially if it is a cheap one or the components are out of alignment.

Finally, the different sensitivity of the human eye to different colours also affects our ability to detect colour in faint stars.

INTERLUDE

Orion the Hunter: An Apparent Irregularity

From February onwards, few who are outside at the right time of evening can fail to note the striking sight of, in Tennyson's line, 'great Orion sloping slowly to the West'. If an observer watches the constellation carefully from early in the year, it is easy to get the impression that something slightly peculiar is happening.

From November to mid January, Orion, observed at the same time each evening, is seen to move steadily westward. After mid January, however, an irregularity seems to arise. The constellation appears to slow down and to hang high in the sky at twilight night after night for an abnormally long time before obviously beginning to sink towards the western horizon.

What is happening is that twilight occurs appreciably earlier on successive nights at this time of year, so that very little apparent movement of the constellation takes place between successive twilights. If position is related to strict clock time, no anomaly arises.

Because the effect of advancing time of twilight gives so strong an impression of the constellation slowing its westward march, it is worth making the effort to note the changing aspect of Orion in its rising and setting cycle from its first appearance in the evening sky in November to its final disappearance in May.

The constellation rises almost vertically, Rigel first and Betelgeuse later by 1 hour 34 minutes. Starting when Rigel is first seen on the eastern horizon at 8.12 p.m. on 4 November, we find that the constellation takes eighty-nine days to reach the meridian (the north-south line through the observer's zenith) at 8.50 p.m. on 1 February, but then takes 142 days to reach the western horizon at 6 p.m. on 23 June. It then sets almost horizontally, Betelgeuse first and Rigel only fifteen minutes later. Thus, twenty minutes after the first star rises, most of the constellation is still below the horizon, but twenty minutes before the last star sets, almost the whole constellation is still visible.

PART 2

Observing with the Star Charts

In the sections that follow, six charts are given, one for each two-monthly period of the year. When you come to use them, there will be one or two practical problems, but none that cannot be overcome with a little effort (don't expect to solve them all in a single night). Comparatively few objects are shown on the charts, especially in the richer regions of the sky. The emphasis is on bright stars, distinctive groups and interesting and important objects.

The first problem will be what you can see from a particular site on a particular occasion. If you live in a large city, its lights will prevent you from seeing all but the brightest stars. This will simplify finding particular areas, but will make faint objects difficult to pick up.

Those who live far from city lights will find the sky confusingly crowded and may have difficulty in separating even groups as bright and distinctive as the Southern Cross from the rich star fields that surround them. After only a little practice, however, the eye will learn to distinguish grades of brightness and the main stars of a constellation will begin to stand out. Remember, in that band of the sky where planets may be found—*planets may be found!* The presence of a bright planet can drastically change the appearance of a familiar constellation, and if it's mistaken for a star it will obviously be impossible to find the expected objects around it.

One of the striking peculiarities of the sky is the number of star groups of similar shape, often close together but of different size and brightness. It is easy even for an experienced observer to note a distinctive group on a chart, then find a similar group in the sky, only to discover later that it wasn't, after all, the required group. With experience, the observer learns to relate size on the chart to size in the sky and is fooled less often.

Those familiar with the northern sky will be able to see many familiar constellations from the south, but they will appear upside-down.

Field Indicator

If you are using either binoculars or a telescope and have provided yourself with star charts more elaborate than those in this book, here is a gadget you will find most useful. Take a small square of light acetate and cut out of the centre a circle equal to the field of your instrument in terms of the scale of the chart. As a guide, the fractionally over seven degree field of 7 × 50 binoculars comfortably spans the Pointers to the Southern Cross (Alpha and Beta Centauri). If you have a telescope, you will require one such mask for each eyepiece. You will find this simple tool almost endlessly useful to show you on the chart how much of a star field you will actually see in the instrument. Make it before you start serious observing; one of the most frustrating things that can happen is to be unable to find in telescope or binoculars a star group that should be unmistakable, then to realise later that not all of it could be seen in the instrument's field at once.

Lastly, remember what was said about 'seeing conditions' early in the book. You can't see faint objects in 'bad seeing'.

It is a good idea to keep a record of observations in a notebook. Record date and time, seeing conditions on a scale of 1 to 10 (1 is a glimpse through haze, 10 is once-in-a-lifetime perfection where most of us do our viewing), the objects observed and their general appearance. Make sketches where possible. Make the notes full enough to tell you what it was all about when you observe the same area again a year or more later. This record-keeping can be quite simple, but is good preparation should you later become interested in serious observing; meticulous records will then be a must.

Ideally, before you begin to observe you should give your eyes twenty minutes to become fully accustomed to the faint light. It is surprising how much difference this makes. Star light alone is quite sufficient to see your way tramping across open country once your eyes have properly dark-adapted. Using an ordinary torch to examine the charts will destroy this dark adaptation. If, however, you wrap several layers of red cellophane around the head of the torch until the light is so dim that you can just comcortably read the charts *after* your eyes have dark-adapted, your night vision will not be affected.

To avoid getting the book damp in dew or otherwise damaged, you can make tracings of the charts. These copies can then have notes scribbled on them, the changing positions of planets marked, comets tracked and unmarked objects inserted all without damaging the book. Read through the descriptive material for the chart you are using and make sure you have a note of every object you wish to observe.

Reading the Star Charts

The charts that follow show the sky for the central part of New Zealand at 10 p.m. at the beginning of each two-month period. Later in the night the most westerly objects will have set and new ones will have appeared above the eastern horizon; the objects rising in the east can be found on the charts for the succeeding months. Observers in the far north and south of the country will find slight variations, but these will not be significant.

The times given can be regarded as becoming earlier by 1 hour every 15 days: that is, a chart which corresponds to the appearance of the sky at 10 p.m. on the first of the month will be correct at 9 p.m. on the fifteenth and 8 p.m. by the thirtieth.

To use the map, hold it so that the horizon you are facing is at the bottom; the lower half of the chart then portrays the half of the sky you are facing. Remember that what you will actually be able to see on any given occasion depends upon ground lights, seeing conditions and dark adaptation of the eyes. Not all objects referred to in the text are marked on the charts. For instance, a small cluster close to a bright star may have been left off to avoid cluttering the chart.

The Milky Way is shown by dotted lines enclosing its brightest parts; again, prevailing conditions will determine how much of it is visible on a particular occasion.

The charts are drawn for the unaided eye, showing third magnitude objects in all areas and fourth magnitude in some. Dot size is an approximate index of brightness. The use of any instrument will reveal a profusion of fainter objects. Galaxies, nebulae and star clusters are marked with an asterisk and their 'Messier' catalogue number where appropriate.

Remember that planets are not marked on the charts and that the appearance of a bright planet in one of the constellations of the Zodiac can radically alter the appearance of the constellation.

With each of the six charts is a broad description of the sky at the times for which the chart is applicable and a description in detail for those areas which are best situated for viewing in that two-month period. Objects which are already well down toward the west will have been described with the chart for the preceding two months, and those objects still low in the east will be covered in the chart for the succeeding period.

When using the charts for the first time, pick a group which is quite unmistakable, such as Orion in summer or Scorpius in winter, and work out from it, making sure of the identity of each new group by reference to the objects around it before going on to more distant groups.

Now wait for a clear night, select whatever chart is appropriate to the time of year, wrap up warmly if necessary (and it usually is, even in summer) and go outside and begin your study of the sky.

Useful Measurements

Where distances are given in this book or taken from an atlas chart or other source, the scale in the following table will be useful in determining roughly the distance on the celestial sphere. Go outside and practise as many of these handy rule-of-thumb measurements as you can. Refer to the star charts that follow for your month and try to locate the stars used. Not all of these guideposts can be seen throughout the year. Alpha Orionis and Alpha Tauri are summer stars. Those in Crux and Centaurus are circumpolar, but will be low to the south in the early part of the night in summer. The charts will show which are accessible at any given time.

Degrees	Example
0.5	Diameter of full moon.
1	Two full moons; tip of index finger at arm's length.
3	Knuckles of first and second fingers at arm's length.
4	Alpha to Beta Crucis.
4.5	Alpha to Beta Centauri; Alpha to Beta Geminorum.
5.5	Long shaft of Southern Cross (Alpha to Gamma Crucis).
7	Tips of four fingers at arm's length.
8	Knuckles of four fingers at arm's length.
11	Alpha Crucis to Beta Centauri.
15	Alpha Crucis to Alpha Centauri.
17	Alpha to Beta Orionis.
20	Alpha Orionis to Alpha Tauri.

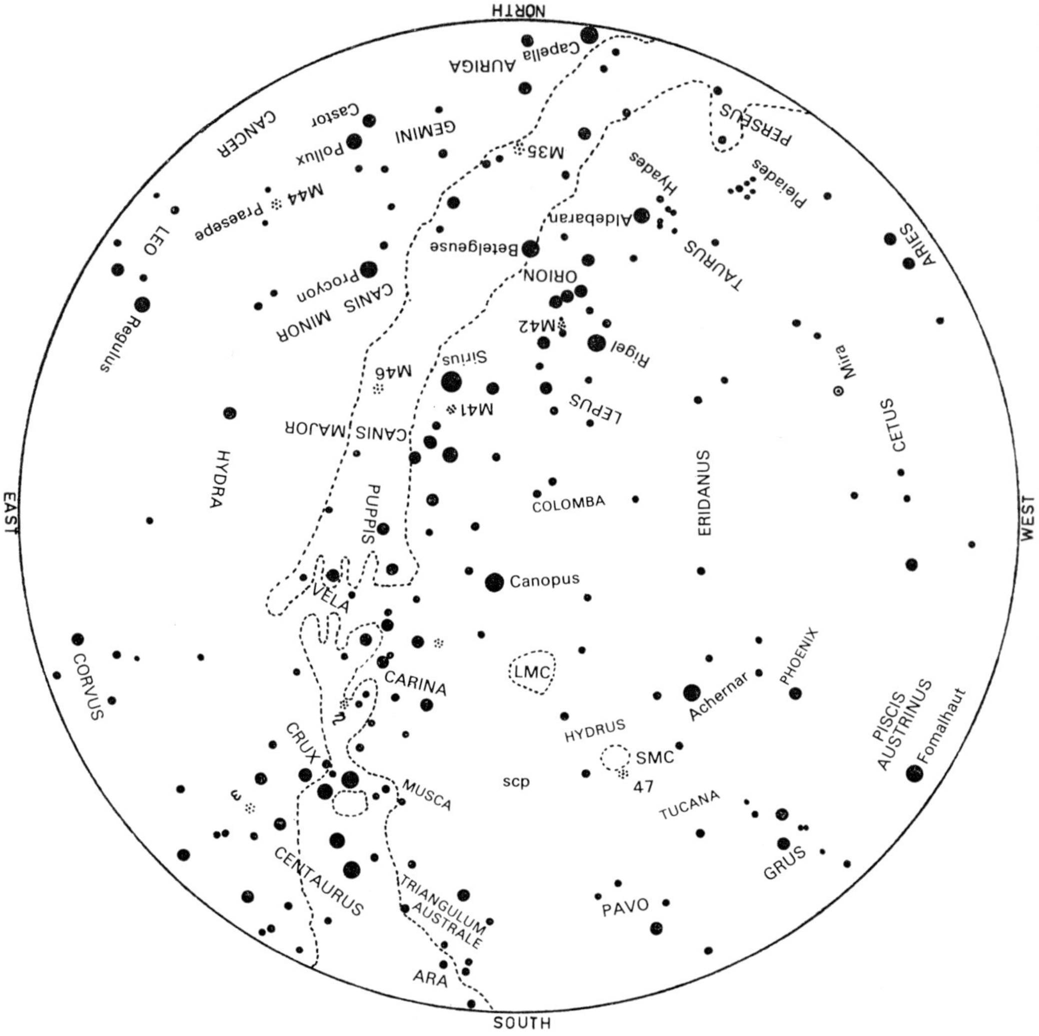

The Sky from New Zealand: **February 1 at 10 p.m.** **March 1 at 7 p.m.**
February 15 at 9 p.m. **March 15 at 6 p.m.**
February 28 at 8 p.m. **March 30 at 5 p.m.**

THE SKY ABOVE US: FEBRUARY-MARCH

In February and March the two brightest stars in the sky, Alpha Canis Majoris (Sirius) and Alpha Argus (Canopus) are almost overhead, the Southern Cross and Pointers are low to the south, the rather nondescript constellations of Corvus (a useful reference group) and Hydra are rising due east, but in the north-east a much richer region is rising. Here Leo is marked by Alpha Leonis (Regulus) and further to the north is Cancer and then Gemini, with Alpha and Beta Geminorum (Castor and Pollux, the Twins) and above them Canis Minor, with its brightest star, Procyon.

Due north, Alpha Aurigae (Capella) is on the horizon (probably invisible except from northern regions) and high in the sky above it is Orion, while to the north-west the two great clusters in Taurus, the Hyades and Pleiades, are setting, the former well marked by the red giant Alpha Tauri (Aldebaran).

In the west, the undistinguished constellation Cetus has almost set, and further around to the south Alpha Pisci Austrini (Fomalhaut) is just above the horizon. Grus and Pavo are low in the south-west with the Small Magellanic Cloud above them; the Large Magellanic Cloud is below Canopus, about a third of the way from Canopus to the South Celestial Pole.

Locating the South Celestial Pole

Near Rigel in Orion a stream of stars begins which flows out above Taurus and down and across the western sky toward Cetus, then up to the star Achernar, near the Small Magellanic Cloud. This is the constellation Eridanus, the River. If you trace the line from Achernar to Beta Centauri (the Pointer nearest the Southern Cross) you will find the Celestial Pole near its centre, where the line extending the shaft of the Southern Cross intersects it. This is the system for locating the South Celestial Pole that scouts and tramping club groups are often taught, but in the author's experience both as an amateur astronomer and a tramper, it is better avoided. In bush, a river gorge or under a partially clouded sky the required stars may not be visible. Even if they are, an observer not sufficiently familiar with the sky may be mistaken in his identification with inconvenient, and possibly disastrous results. It is far safer to spend a little time determining the axis of the arc on which a star or group of stars is moving. Facing south, objects will be circling around the Celestial Pole east to west. The location of the pole can then be estimated quite simply and with far less chance of error.

High above Achernar, near the zenith, is Canopus, relatively close to Sirius in the sky. It is a star of supernal brilliance 110 light-years away; this comparison tells us quite a lot about the two stars, for Sirius appears to us to be the brightest star in the sky, but is only 8.7 light-years distant. Two-thirds of the way from Canopus to the star above the 'E' of VELA on the chart is the star Gamma Velorum, at the end of the largest lobe of the Milky Way. Two companions are of sixth and eighth magnitude, the former being a useful indication of the faintest star visible to the unaided eye in good conditions. To see the latter, of course, you will need an instrument. A small telescope will show several more stars close by of ninth magnitude, making a most interesting group.

The Milky Way is our view, from our position in the plane of our galaxy, of that dense disc of stars, gas and dust which makes up the central part of the galaxy. Looking outward in this plane we see many stars, but especially the combined light of the many millions that lie too far away to be seen singly. Conversely, when we look out at right angles to this plane we see few stars, for in such regions we are looking through the spherical volume of the galaxy which is only thinly populated with stars.

Strange Rogue Star

Half way along the line joining the 'U' of CRUX to the 'C' of CARINA on the chart is an asterisk marking one of the most remarkable objects in the sky, the 'rogue' star Eta Carinae. When first catalogued, it was bright enough to be given a Greek letter; later it became even brighter. Now it is just below unaided eye visibility. It is a distinctive yellow colour in the telescope and is fuzzy in appearance. This comes from the compact cloud of glowing gas and dust that surrounds it. Near by is another bright patch of nebula with a distinctive keyhole-shaped dark patch on it, although a small instrument may not show this.

Eta Carinae is truly strange; indeed, one astronomer has said that it may not even be a star as we ordinarily understand that term. In 1843 it blazed up to rival Sirius, and stayed that way for ten years. After thirty-three years of gradual fading, it passed below unaided eye visibility and has never risen above it since, although it does fluctuate slightly. This is one of the objects that astronomers are moving south specifically to study. Many of its characteristics are flatly contradictory. It is an extremely powerful source of low-energy infrared radiation, which probably originates in a dust-filled volume of space which may be very extensive. At the centre of this volume, however, is a source of immensely powerful high-energy radiation. No explanation has been devised which fully explains the processes going on in this strange star.

Most astronomers, amateur and professional, glance at this region of the sky whenever they are outdoors on a clear night. Eventually Eta Carinae is almost certain to come blazing up to unaided eye visibility again.

The nebula itself is well worth study. Especially in binoculars, the whole region is rich in pairs of stars, nebulae and small clusters, all objects in one of the densely populated spiral arms of our galaxy.

To the north, right on the horizon, is the northern constellation, Auriga, with its brightest star Capella. It is worth going to some trouble to glimpse this star, which is the third brightest in the northern sky. It is actually a double star, both components yellow-white like the Sun, but bigger and brighter. The fainter of the pair has a more yellow tinge. It is 45 light-years distant.

This is a good time of the year to compare the famous northern and southern pairs, the Twins, Castor and Pollux, and the Pointers, Alpha and Beta Centauri. Both pairs are at about the same altitude on opposite sides of the sky.

Castor and Pollux are slightly closer, but the Pointers are the brighter pair. Alpha and Beta Centauri are the third and tenth brightest stars in the sky; Pollux (Beta Geminorum) is seventeenth brightest, and Castor, fainter despite its designation of Alpha Geminorum, is much farther down the list of bright stars. Taking the Pointers plus the four brightest stars of the Southern Cross, there is no other equal area of sky which contains six stars so bright. Indeed, of the twenty-five brightest stars in the sky, more

THE REGION of the unusual object Eta Carinae (above) and the nebulosity surrounding it (right). This region is close to the Southern Cross, on the opposite side to the Pointers and in a corresponding position. Note the delicate traceries among the gas clouds and the numerous galactic clusters among the surrounding stars—a typical view among the star clouds of one of the brightest parts of the Milky Way. *(South African Astronomical Observatory; Uppsala Southern Station)*

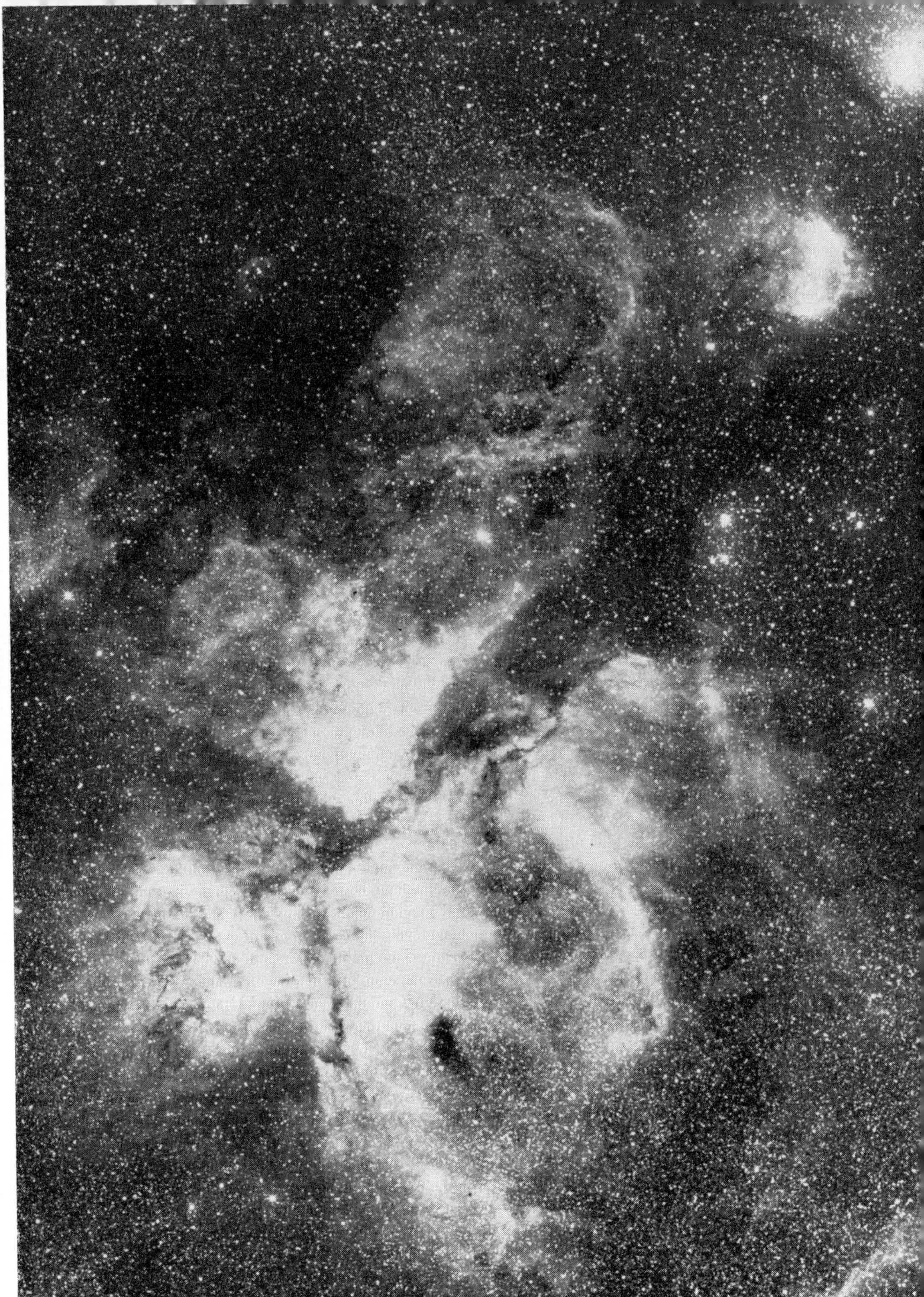

than half, thirteen, are south of the Celestial Equator, although it is as well to remember that they include stars, such as Sirius and Rigel, that are close enough to the Celestial Equator to belong to what we loosely think of as 'northern' constellations.

This heavy weighting of the brightest stars in favour of the southern sky is one of the effects of the rich central regions of the galaxy, the brightest part of the Milky Way, being in the southern sky. This is not all advantage: Castor and Pollux, despite being the fainter pair, do tend to look rather more impressive than the Pointers because the Twins are in a relatively dark area of sky; the Pointers are in an exceptionally rich area.

Castor and Alpha Centauri both illustrate an important point. We tend to think of stars as being single orbs, like our prototype star, the Sun. Even a small telescope will show that both these stars are double (binoculars will not). The single object visible to the unaided eye is actually two stars close together in space and revolving around their centre of mass.

The two components of Alpha Centauri are stars much like the Sun, especially the brighter one. They are separated by about the distance between the Sun and Uranus, and take about as long (80 years) to orbit each other. There is a third component of the system, a faint red dwarf very distant from the two bright stars. This dwarf was long believed to be the nearest star of all to the Sun and accordingly was named Proxima Centauri. Modern measurements put all three at the same distance. Proxima is a 'flare star', that is, at irregular intervals it undergoes a rapid but short-lived increase in brightness. There is no corresponding temperature rise and nobody yet knows what causes these mysterious flares.

In the case of Castor, each of the two components visible in a small telescope is itself a double star; each pair is very close and the fact that there are two stars can be detected only by sophisticated instruments.

Several Bright Clusters

While working in this part of the sky, look for the several clusters marked by asterisks on the chart. In Gemini, M35 is a fine object, and so is M44 in Cancer. Two more, M41 and M46, flank Sirius.

ALPHA (left) and **BETA CENTAURI**, the Pointers to the Southern Cross. The star at top right is Epsilon Centauri. To the unaided eye or in binoculars the two stars appear identical, but in actuality Alpha is double and relatively close and Beta single and much further away. They are a more spectacular pair than the famous northern twins, Alpha and Beta Geminorum (Castor and Pollux) but the southern pair suffer from the competition of the surrounding bright region of the Milky Way. *(South African Astronomical Observatory)*

Objects of a different kind are M1 and M42 which we will discuss later.

The 'M' designation, incidentally, stands for 'Messier', the name of a French astronomer (1730-1817) whose great ambition was to discover comets, and in fact he found more than a dozen. To aid his searches, he compiled a list of 109 objects which could be confused with comets. He was working with a small telescope, which is why the list includes a number of objects which couldn't possibly be confused with a comet when seen in a large telescope. Two of his 109 objects cannot now be found and were, presumably, comets which he failed to recognise. Charles Messier's irony is that he is remembered for his list of non-comets, not his real ones.

M44, also known as the Praesepe Cluster, was revealed for what it is by Galileo. In his day this object was termed a nebula from its cloud-like appearance and it was one of his early discoveries that it was seen to be a cluster of stars when viewed in a telescope. It is popularly known as the 'Beehive', or 'Manger'.

The Celestial Dogs

The Big and Little Dogs, Canis Major and Canis Minor, follow Orion along the Celestial Equator. Alpha Canis Minoris is Procyon, a double star, but the companion, at magnitude eleven, is far too faint to concern us. Alpha Canis Majoris (Sirius) also has a faint companion, and in this case a very famous one: it is the prototype white dwarf, a state referred to earlier in the discussion of the eventual decline and death of the Sun. It is a star in which the hydrogen fuel has been almost entirely used up and in which radiation pressure will no longer support the outer layers. It has collapsed to small size and immense density. Sirius' companion was the first such star to be discovered.

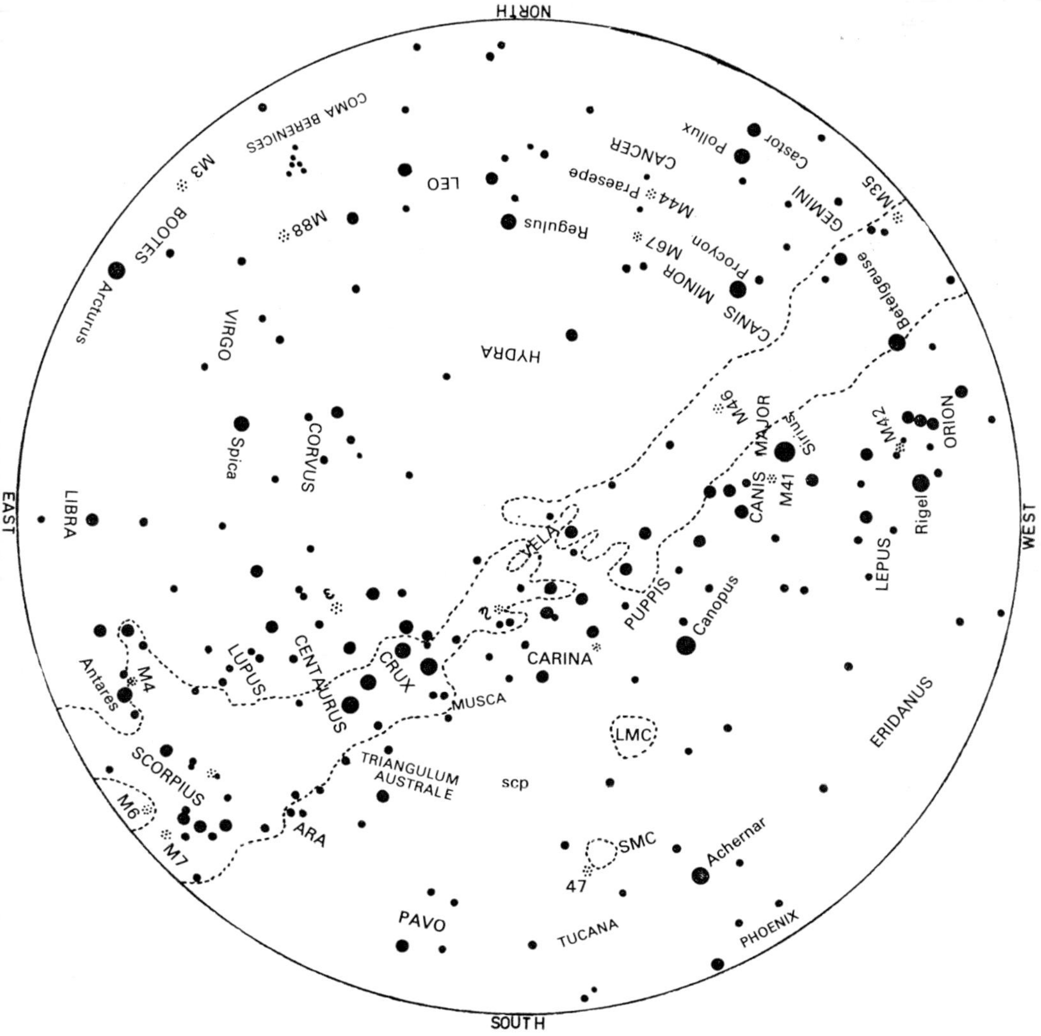

The Sky from New Zealand:

The Sky from New Zealand:	April 1 at 10 p.m.	May 1 at 7 p.m.
	April 15 at 9 p.m.	May 15 at 6 p.m.
	April 30 at 8 p.m.	May 30 at 5 p.m.

THE SKY ABOVE US: APRIL-MAY

The Centaurus, Crux and Carina regions are now much higher in the sky, while to the south the Magellanic Clouds are sinking lower on the western side of the Celestial Pole. Phoenix, Tucana and Pavo are right-to-left across the southern horizon. In the east, Scorpius is rising with Lupus above it, Libra is due east and Bootes further north. Virgo and Corvus are high enough to be worth viewing. To the north, Leo and Cancer are higher than on the previous chart, with Hydra above them. Gemini is now setting in the north-west and Orion is well down, although the Big and Little Dogs above it are still high enough to view. Puppis and Vela are near the zenith, rich in bright star clouds. Lepus and Eridanus are setting in the south-west. The Milky Way stretches from north-west to south-east, passing virtually through our zenith.

Low on the northern horizon it may be just possible to glimpse some of the most southerly stars of the famous northern constellation Ursa Major, the Great Bear, although the figure of the bear itself, with its most un-bear-like tail, is far below our horizon.

The constellation Leo is as high to the north as we ever see it. Alpha Leonis is Regulus, by far the brightest star in the group and quite a famous one. Hipparchus, the Greek astronomer who lived about 140 B.C., observed it carefully enough to note that there was apparent movement, the first observation ever made of the apparent changes in the sky caused by the wobble in the Earth's axis of rotation, correctly known as the 'precession of the equinoxes' (this phrase actually describes the astronomical consequences of the wobble, rather than the cause). Hipparchus is commemorated in the name of a great ring plain, 160 km across, near the centre of the Moon's disc. Several stars in the constellation are double or multiple systems and it repays a careful study, for, being far from the Milky Way, Leo is not rich enough to be cluttered and confusing.

Leo also produces a meteor shower in mid November. It is highly variable from year to year, but at its best produces many bright meteors which have the characteristic of leaving long-lived trails in the atmosphere. It was this shower which produced the spectacular meteors of 1799, 1883 and 1966, the first of which was the subject of a report by the scientist Von Humboldt, in

Venezuela at the time, who saw the sky continuously filled with thousands of meteors in the early part of the shower. Humboldt's name was conferred on one of the great ocean currents and it is in connection with oceanography that he is best remembered today, but he saw in 1799 one of the greatest sights the sky affords.

The Southern Cross

The Southern Cross is now high enough to be worth examining in detail. It is a good example, incidentally, of the changing position of the sky relative to the Earth caused by the precession of the equinoxes. The Cross was not always so southerly a group. Two thousand years ago it rose above the southern horizon of Palestine and if Christ had looked south from the hills above Galilee in June or July he may have seen its distinctive form low on the horizon. It is a spectacular group, and has always been noted by whole peoples and individuals far enough south to see it.

Strictly speaking, it is not a perfect cross; shaft and arms are at something less than a right angle. But Alpha Crucis is a first magnitude star, Beta and Gamma second magnitude and Delta third, and not many close groups of four are so bright.

Alpha Crucis is a system which a small instrument (63 mm) will resolve into a double comprised of bright and faint components, but a slightly larger telescope will show the bright component itself to be double. They are moving in the same direction at the same speed, but must be much further apart than the components of Alpha Centauri for there is little detectable movement relative to each other.

Close to Beta Crucis, on the side of the Cross toward the Pointers, is a small cluster centred on the red star Kappa Crucis. It is notable for the wide range of colours in its stars, although it needs a large telescope to show this to advantage. John Herschel called it the 'Jewel Box', which describes it well.

Between the Cross and the Pointers lies an area of the Milky Way much less rich than those surrounding it. It is necessary to get well away from ground lights and to have a truly dark, clear sky before it can be seen. This is the famous Coal Sack, a dark nebula which appears so big because it is relatively close. It is

made up of gas and dust like the bright nebulae we have already examined around Eta Carinae, but there are no bright stars within to illuminate it.

Two things about the Coal Sack should be realised. First, it is not particularly dense. It achieves its obscuring effect because the dust is very fine and so produces a highly efficient 'damping' effect, and because it is many light-years thick. Secondly, this region is not devoid of stars, and the stars beyond the nebula are not entirely obscured. Some of the stars we see within its boundaries are probably between us and its densest regions and the light of stars beyond it is only dimmed.

While working at Gisborne, the author once made a trip through to the mouth of the Motu River in the southern Bay of Plenty to pick up a party making a trip down the river by rubber dinghy. They didn't arrive on the first evening and a night was spent camped on a shingle bank about five kilometres from the mouth. The night was crystal clear, the nearest ground lights were Opotiki, Whakatane and Taneatua and it was possible to see that the Coal Sack contained a number of faint stars and two or three brighter

KAPPA CRUCIS CLUSTER (above left), close to the Southern Cross. This is the Jewel Box Cluster, a host of coloured stars set around Kappa Crucis. In good conditions and viewed with a reflecting telescope (the greater the aperture the better, but at least 150 mm) it can look like a handful of precious stones scattered on black velvet around the red ruby of Kappa, but in poor conditions, a small instrument or most refractors, it will disappoint. This plate displays the 'spikes' often seen on bright star images which result from the supports of secondary mirrors or plate-holders in reflecting telescopes. The plates taken with lens systems (refractors) are free of these effects.

(Radcliffe Observatory)

SOUTHERN CROSS AND COAL SACK NEBULA (below left), a cloud of dark gas which dims and obscures many of the stars beyond it. Note that the relative brightness of the stars of the Southern Cross is different on the photographic emulsion from what the eye sees due to the different light sensitivity of the emulsion. Close to Beta Crucis and the edge of the Coal Sack can be seen the tight little knot of stars which is the galactic cluster known both as the Kappa Crucis Cluster (from its brightest star) and the Jewel Box. Beta Centauri, one of the Pointers, is at the left edge of the plate and at the top left is Epsilon Centauri. The line from Beta through Epsilon, if extended almost as far again, leads to the globular cluster Omega Centauri.

(South African Astronomical Observatory)

ones. To eyes accustomed to the effects of city lights, even the form of the Cross was almost obscured by the profusion of stars in and around it. It was a magnificent sight.

The Cross is wholly surrounded by Centaurus, and indeed formed part of that constellation until the fifteenth century. Centaurus is one of the glories of the southern sky and contains a number of interesting and important objects. A binocular search for double stars and small clusters is particularly rewarding, for they abound. Several of the bright 'Greek letter' stars, like Alpha, are doubles, although not many can be resolved by binoculars.

About mid-way along the line from Alpha Crucis to Eta Carinae is Lambda Centauri which marks a striking field and the cluster 47 Centauri, also sometimes given its New General Catalogue number, NGC 3766. This is actually a star cloud of considerable size formed by a great number of stars, but dimmed and shrunk by distance to the glistening little object we see in binoculars.

The Globular Clusters

The Milky Way is particularly bright in this region of the southern sky, because here we are looking in a direction which leads not only through the width of the dense central disc of the galaxy, but coincides also with the spiral arm on the edge of which the Sun is situated. We will look now for an object which lies far out from the central plane and spiral arms, one of the clusters whose realm is the spherical volume of thinly populated space which comprises the bulk of our galaxy.

Find on the chart the star which forms an equilateral triangle with Beta Centauri and Beta Crucis. This is Epsilon Centauri. Now take the line from Beta to Epsilon Centauri and extend it approximately the same distance. This will bring you to an obviously hazy, fourth-magnitude patch of light, marked on the chart with the Greek letter Omega (like the small letter 'w'). Binoculars show it clearly. Here is the brightest of the globular clusters, spectacular spherical star systems containing many hundreds of thousands of stars. A few are closer, but they are either smaller or are partially obscured by nearer objects, or both.

Omega Centauri is the most spectacular of these magnificent objects. A telescope of about 20 cm aperture will resolve most of

OMEGA CENTAURI, closest and brightest (magnitude 4.5) of the globular clusters. Note the spherical shape and the increasing density of stars toward the centre. Many of the stars on the plate are between Earth and the cluster, but although the cluster clearly lies almost entirely inside a circle 7 cm in diameter on the plate, many stars more distant from the centre are part of it. The globulars belong to the 'halo' of the galaxy and Omega Centauri is far from the central plane that contains the spiral arms and Earth's Sun. Note the distinctive form of the globular clusters and how they differ from the open clusters of the spiral arms, such as the Pleiades and Kappa Crucis.

(South African Astronomical Observatory)

the cluster into stars. We can seldom view it in good conditions from Wanganui, but when we can, it shares with Saturn the distinction of being able to raise gasps of amazement on public viewing nights. The stars individually visible in the cluster in a small instrument are only the giants of the system. Between them are many smaller, fainter stars we cannot see and when it is realised that this cluster is 22,000 light-years away we begin to have some appreciation of what an incredible object it is.

About 100 globular clusters are known and the existence of more in the outer reaches of the galaxy, screened from sight by the nucleus and disc, can be deduced, but nowhere else in the sky is one so clearly seen as Omega Centauri. These clusters have become important objects of study, not merely for what they tell us about some of the more distant parts of the galaxy, but also because, as is true of all clusters, the stars comprising them have a common composition and a common age. The differences between them are therefore the differences that arise from mass alone.

Incidentally, many authors of fact and fiction have fallen into error in speculating about what the inhabitants of a planet circling a star inside a globular cluster would see. In spite of the vivid description in Isaac Asimov's 'Nightfall', probably one of the most famous science fiction stories ever written, it is unlikely that any star other than the planet's primary would be close enough to show a disc. The night sky of such a planet, however, would contain several hundred stars as bright or brighter than the first-magnitude stars in Earth's sky, and distributed evenly all over it. To the inhabitants of such a planet, the cluster would appear to be the whole universe and they would be able to detect little of the galaxy beyond it, at least at optical wavelengths. Radio astronomy might tell them rather more. Asimov, of course, would have been well aware of all this and was only using author's licence (read *Asimov on Astronomy* some time; it is one of the most entertaining of his factual books).

NGC 5128—the Freak

Close to Omega Centauri is a faint patch of light worth searching for. Good binoculars will show it (it will be in the same field as

NGC 5128 in Centaurus, the abnormal galaxy which is also known as Centaurus A, one of the most powerful radio sources in the sky. Not far from the Southern Cross, it can be found, in good conditions, in the same 7 x 50 binocular field as the globular cluster Omega Centauri. The Omega cluster is part of our galaxy, but NGC 5128 is much further away. It is a star system comparable to our galaxy, but with unusual features. Many galaxies which show abnormal characteristics visually are also powerful radio sources, and are not yet fully understood. Photographed through the 5 m Mount Palomar telescope. *(Hale Observatories)*

Omega Centauri if 7 $\times$ 50s are being used), but it is faint and elusive and will be visible only under the best conditions. With binoculars you will see only a featureless patch of light but a bigger instrument will show a circular, hazy patch with a dark band across its centre.

This object does not radiate very much visual energy, but when a radio telescope is pointed at it the effect is very different. This is the galaxy referred to in radio astronomy as the Centaurus A source, one of the most powerful radio objects in the sky, but here we will use its visual designation, NGC 5128. In the early 1960s the famous astronomer Harlow Shapley described this object as a freak, the nature of which was one of the many problems posed by the galaxies.

It is at least 15 million light-years away and if it were anything less than a large galaxy, it obviously would not radiate enough light energy to be seen at such a distance. What, then, could cause it to be so much more powerful at radio wavelengths? An early conclusion was that it was not one galaxy, but two—and that they were in head-on collision. This, however, is no longer thought to be likely. There is only one nucleus and NGC 5128 is a good example of the general rule that many galaxies which are visually abnormal are also strong radio sources. It is, in fact, a fairly typical radio galaxy, although such objects are not yet well understood.

To the north and east at this time of the year is Hydra, a long, trailing constellation appropriately representing a sea serpent. It contains little of interest, but there are a number of faint nebulae and clusters for the binocular, and more especially the small telescope, observer. The irregular variable U Hydrae drops a full magnitude from 4.8 to 5.8, so is always visible to the unaided eye in good conditions.

Corvus contains nothing of note at the level at which we are working, but the small, distinctive group of bright stars is worth getting to know, if only as a guide to objects in the adjoining constellations.

At the foot of the Southern Cross is Musca, on the edge of the Milky Way and rich in faint stars and doubles. Almost between Musca and the South Celestial Pole is Chameleon.

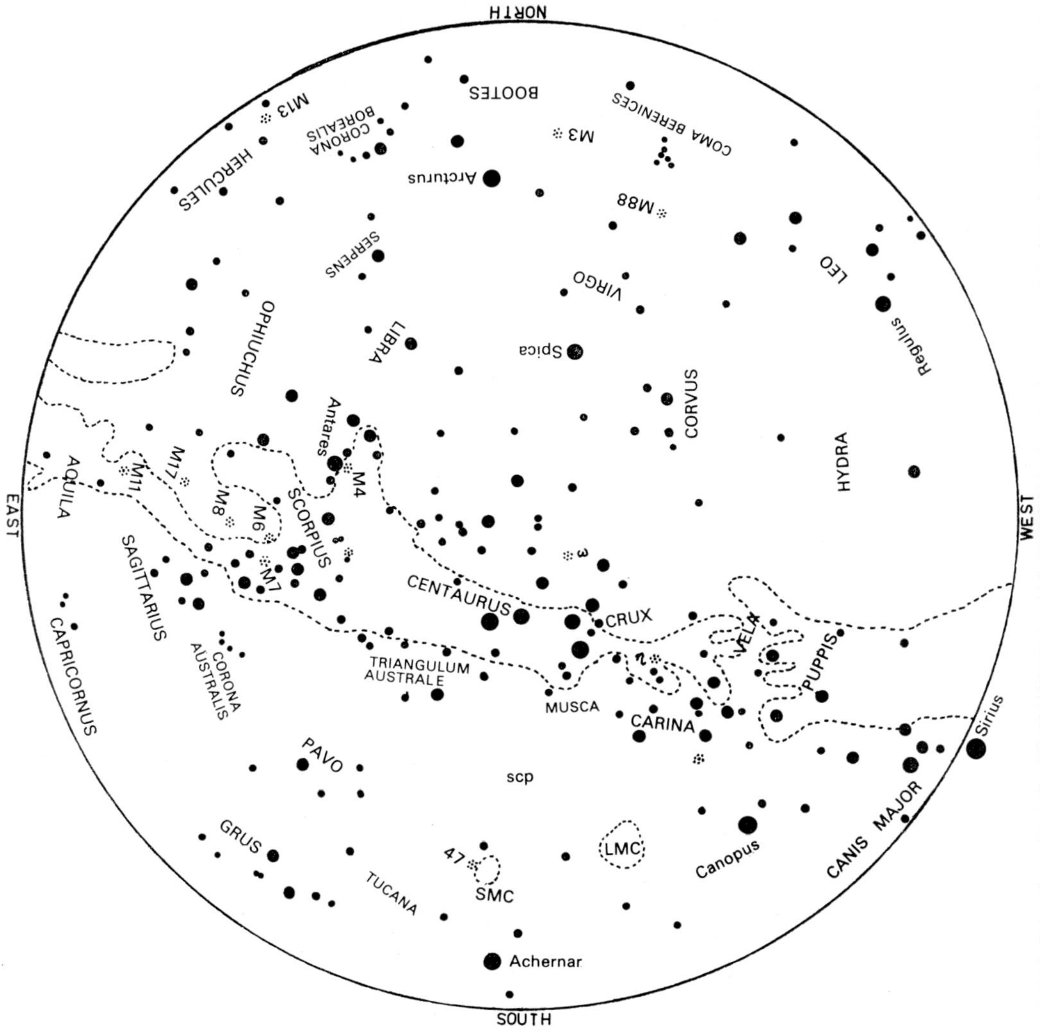

The Sky from New Zealand: June 1 at 10 p.m. · July 1 at 7 p.m.
June 15 at 9 p.m. · July 15 at 6 p.m.
June 30 at 8 p.m. · July 30 at 5 p.m.

THE SKY ABOVE US: JUNE-JULY

The Milky Way stretches east to west across the sky, its brightest part (in Sagittarius) low to the east, then rising through Scorpius, Centaurus and Crux near the zenith and descending through Vela and Puppis to the western horizon. The last stars of Canis Major are sinking below the western horizon, the bright stars Canopus and Achernar are low in the south-west and south, and around to the east we have Tucana, Grus, Pavo, Corona Australis, Capricornus and Aquila. North of the Milky Way and low down is Hercules, Ophiuchus and Serpens above it, with Corona Borealis almost due north, then Bootes and, further west, Coma Berenices. Corvus, Virgo and Libra are high in the northern sky. Leo and Hydra are setting in the north-west.

Alpha Virginis (Spica) is the brightest star to the north, with Alpha Bootis (Arcturus) below it. The object marked M3, actually in Canes Venatici, a constellation not named on the chart, is a globular cluster.

Arcturus is the fourth-brightest star in the sky and the brightest north of the Celestial Equator. It is relatively close, about forty-one light-years away. It is also moving quite rapidly. In our galaxy all objects outside the nucleus are in orbit around the nucleus, the gravitational centre. The orbits are, of course, very large and the movement difficult to detect, mostly because almost everything in our vicinity is moving in the same direction. Arcturus, however, is crossing the plane that contains the spiral arms and so its movement, different from the Sun's, is more easily detected, at least over a period of about 200 years. In the eighteenth century the sixth magnitude star close to it was on the other side. Arcturus is a giant star, more than twenty times the diameter of the Sun, rather more massive, but much less dense.

Spica is a double star, but don't expect any telescope to separate the components: they are very close. It is the sixteenth-brightest in the sky, almost exactly magnitude one. Remember that Mercury takes about 88 days to complete its orbit around the Sun, then reflect that the components of Spica take only 4 days to complete their circuit. Each must raise huge tidal bulges on the other, yet they are probably distant enough not to interact directly, at least until one begins an expansive phase. Among the objects of most intense study in the sky today are binary (double) stars where the components are in virtual contact, grossly distorted stars in which a constant stream of matter is pouring from one to the other and producing strange effects on the evolution of the system.

Galaxy Rotates

If we again visualise our Milky Way Galaxy as a great sphere with a densely populated equatorial plane, we can now add another fact to what we already know about it: it rotates about an axis passing through its nucleus at right angles to the equatorial plane. It does not, of course, rotate as a solid object but, like the solar system, the inner regions most rapidly and the outer parts very slowly. It follows that the galaxy has two poles, analogous to Earth's poles, at either end of its axis of rotation. The North Galactic Pole lies in the constellation of Coma Berenices, whose tight little cluster of stars now lies well down to the horizon below Virgo. The South Galactic Pole is in the constellation Sculptor.

As we look out from inside our galaxy, we have to be careful to sort out what is real and what is only apparent, caused by some aspect of the galaxy. One of the apparent things about the universe as we see it is that galaxies are not usually found near the Milky Way in the sky, but are common in the less-rich areas. This effect isn't real. The galaxies are more or less evenly distributed around us, but the richest parts of our sky obscure them. Obviously, too, the direction of the galactic poles must be at right-angles to the disc of the galaxy in relatively star-free parts of the sky.

As Virgo lies in one of these star-free areas, close to one of the poles of the galaxy, we would expect to be able to see out of our galaxy in this direction to some of the other galaxies in the universe at large. This is exactly what we do find, and the view is enhanced by the fact that in this direction the external galaxies are particularly numerous. Almost any instrumental aid will show, in good conditions, numerous faint patches of light in this constellation, each of which is a galaxy, some like our own, some very different. Some are relatively close, as distances are measured between the galaxies—perhaps only a few million light-years away; others are certainly among the most distant objects a small instrument can

show, many millions of light-years across the void. They are mostly members of a cluster of galaxies: the Virgo Cluster.

Virgo is an area of great interest to astronomers with big telescopes, but less so to amateurs. The area from Spica across to the adjacent constellation of Libra is worth some attention, however,

MILKY WAY, showing one of its brightest parts, from Centaurus (the Pointers are lower right) to Scorpius. The bright star clouds at the top left of the photograph are in and around the curve of stars which form the 'tail' of the Scorpion. This was a two-hour exposure using a lens of only 1.3 cm aperture and 15 cm focal length. *(Royal Astronomical Society)*

for its star fields. If, on a more elaborate chart, Iota Virginis can be identified, it will be found to be at the heart of a small, open cluster.

Coma Berenices is one of those constellations made for binoculars. The stars are too widely spread to be seen in the narrow field of a telescope, but the Y-shaped group of stars shown on the chart fits the field of 7 × 50 binoculars. Seen in a dark Northland sky, it is quite striking, even to the unaided eye.

Libra is one of the constellations of the Zodiac and it is worth noting in passing that it is the only Zodiacal constellation which is an inanimate object, not a living creature, real or mythical. Alpha Librae is a wide double. It is the star just above the 'I' of LIBRA. Beta Librae is the star below the 'B', and about 4 degrees west of Beta (4 degrees is 8 diameters of the full Moon) is Delta Librae, which is sometimes as bright as 4.8 magnitude and sometimes as faint as 6.1. The variation takes place with clockwork regularity over a period of 2.3 days but it is not a variable star in the strict sense. In this case the star is double and the plane of the orbit lies almost in our line of sight. One component is so faint that we cannot see it, but once in each revolution it partly obscures the brighter component. Such stars are known as 'eclipsing variables' of which the prototype is Beta Persei, better known by its Arabic name Algol which means 'Demon Star'. It is thought that this name indicates that the ancient Arabian astronomers knew that it was variable, which may have been the case because its drop from 2.2 to 3.5 magnitudes in 2.87 days is quite marked. Algol is too far north to concern us here, but is mentioned because such stars are often described as 'Algol-type'.

Scorpius is one of the few constellations which bears an obvious resemblance to the creature for which it is named—the great sweep of the tail is quite unmistakable. This is one of the richest regions of the sky.

Alpha Scorpii is the red giant Antares, or 'Rival to Mars', although to make sense of the name it is necessary to know that Ares was the Greek name for the god the Romans called Mars. In this case 'giant' really does mean what it says, for its diameter is some 285 times that of the Sun! Antares has a much fainter green companion which, in theory, a 10 cm telescope will show, but in practice the glare of the bright red primary makes it extremely difficult to see in anything short of excellent viewing conditions. The best way to see it is on the rare occasions that Antares is occulted by the Moon: the companion is first to reappear.

In this extremely rich area clusters abound. Look for M4, a compact globular cluster close to Antares, look for another half way along the stream of stars leading toward the tail, and for M6 and M7 below the tail toward Sagittarius.

Because of the crowding in this area, the charts in this book are unable to do it justice and it is well worth obtaining, even on loan from a library, a star atlas such as Norton's (see bibliography) and tracking down the interesting objects one by one.

The star adjacent to the cluster marked along the Scorpion's back is Zeta Scorpii. The next one back toward Antares is Mu Scorpii. Both are easily seen to be doubles. The two stars right at the end of the tail are Lambda and Upsilon Scorpii and between them and the Mu pair is a rich field for the binocular observer.

Between Scorpius and the zenith is the small constellation Lupus and although it contains no specific object to interest us, it would be a quite striking star group if it were not so completely surrounded by the rich star fields of Centaurus and Scorpius. The same is true of Ara on the other side of the Milky Way.

Below the Pointers, looking south, is Triangulum Australe, the Southern Triangle. It contains two quite bright variable stars of a type known as 'Cepheid' after the prototype, Delta Cephei. This class of variable is important because there is a relationship between the period and the actual brightness (absolute magnitude) of the star which enables the brightness to be determined by measuring the period. From the difference between the absolute magnitude and the apparent magnitude (how bright it appears to us) the distance can be calculated.

Because these are bright stars and are observable out to great distances, they have proved to be a useful measuring device to determine distances not only within our galaxy, but of galaxies beyond its boundaries.

Looking east, the Northern and Southern Crowns, Corona Borealis and Corona Australis, may be compared at approximately equal altitude. This is about the best view we get of Corona Borealis, but Corona Australis will rise much higher in the months to come.

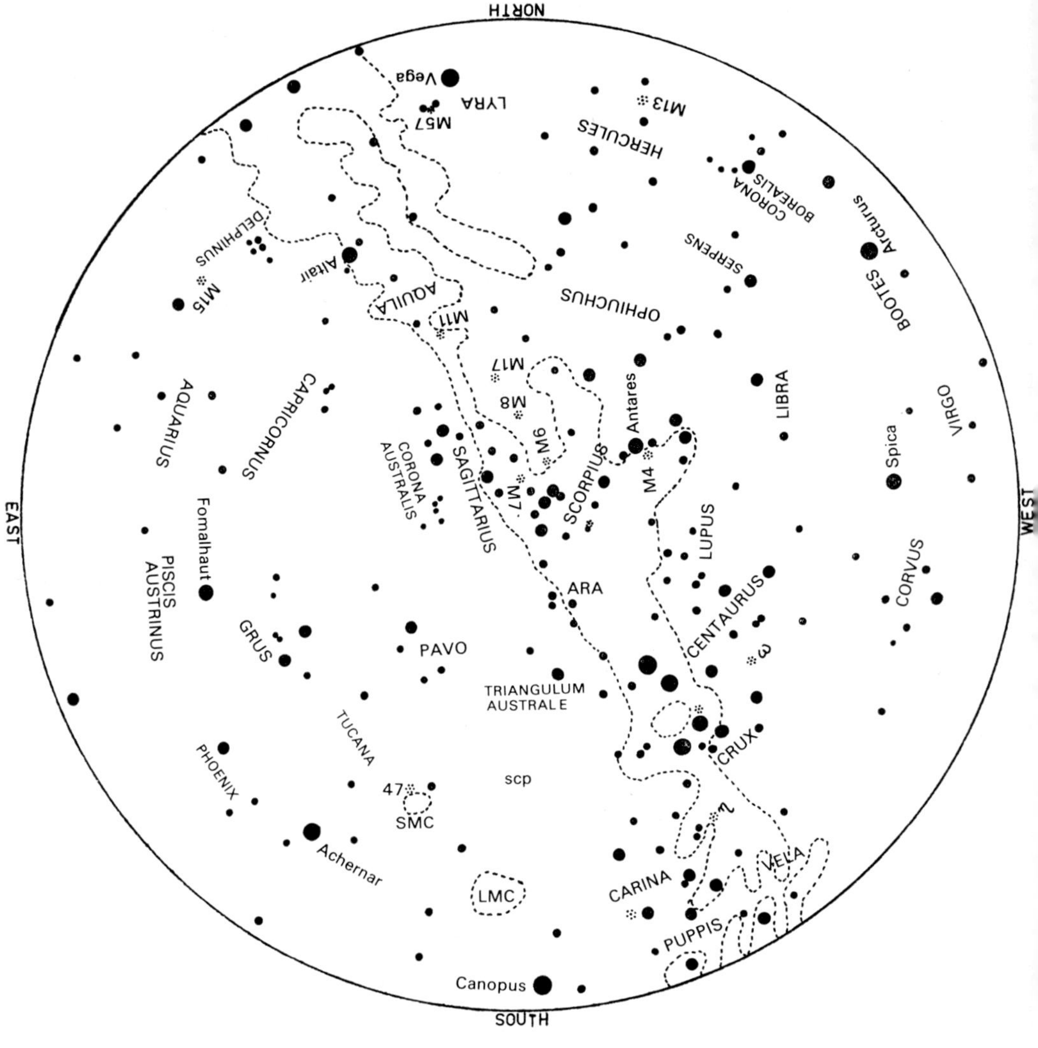

The Sky from New Zealand: August 1 at 10 p.m. | August 15 at 9 p.m. | August 30 at 8 p.m. | September 1 at 7 p.m. | September 15 at 6 p.m. | September 30 at 5 p.m.

THE SKY ABOVE US: AUGUST-SEPTEMBER

The Milky Way still passes near the zenith and it is now that we see to best advantage the brilliant heart of the Milky Way Galaxy, or at least the star clouds which surround its heart, in this magnificent view of Sagittarius. It is a view forever denied northern viewers. From the north-east around to the north-west we have Delphinus, Lyra, Hercules and Corona Borealis, with Aquila and Ophiuchus higher in the sky. In the west Virgo and Corvus are almost gone, and Libra is just above Virgo. Antares and the whole Scorpius region is perceptibly dipping downward. To the south, the Southern Cross is west of the Celestial Pole, the Pointers above it, and Carina and Vela are well down. Canopus is right on the horizon directly below the pole.

From south-east to north-east, Phoenix, Piscis Austrinus and Aquarius are rising, Grus and Capricornus above them. Corona Australis is almost in the zenith.

Having just described Sagittarius as 'magnificent', it is something of a comedown to have to acknowledge that it contains no first-magnitude star—but it doesn't have to. This is the direction toward the heart of the galaxy, and although the nucleus is forever hidden behind the star clouds, dust and glowing gas, we can see enough to grasp the glorious vistas in this direction. The Sun is about 27,000 light-years from the nucleus; we can see perhaps two-thirds of that distance through the Sagittarius star clouds.

Alpha and Beta Sagittarii are the two brightest stars of the constellation. Beta is a double wide enough for both components to be separately visible to the unaided eye, and Gamma, deep in the bright part of the Milky Way, is also a wide double. The object marked M8 on the chart is a cluster just visible to the unaided eye, and just north of it the patch of luminosity at magnitude nine and clearly visible in binoculars is the famous Trifid Nebula, photographs of which appear in most books on astronomy. Don't expect, of course, to see with binoculars what long exposures with some of the world's best telescopes show, but famous objects are always interesting, even in small instruments.

Like Scorpius, this region is far too rich to be adequately shown on the chart or described. It will provide endless enjoyment if searched carefully and systematically on a clear night, particularly with binoculars which will show its splendid star fields to best advantage. More elaborate charts will permit detailed identifications.

Before leaving the region, pick up the regular curve of stars just at the limit of unaided eye visibility: the constellation Corona Australis. It isn't spectacular, but it is useful to be able to identify the Southern Crown.

Low to the north, just east of the Northern Crown, is the globular cluster M13, the best of this class of object visible in the Northern Hemisphere. It does not look as poor to northern viewers as it does to us because they see it much higher in the sky than we ever do, but it certainly isn't the equal of Omega Centauri.

Above M13 in Hercules is Ophiuchus, the forgotten thirteenth constellation of the Zodiac for which there is no sign. The name derives from Aesculapius, God of Healing. It is a big constellation (the planets take longer to cross it than Scorpius) but not a spectacular one. Where the branch of the Milky Way stretches into it from the direction of Scorpius are some attractive fields with a group of five stars, each of which is double. The brightest of these is Theta Ophiuchi. There are six globular clusters worth looking for, although none is bright.

Aquila, in the Milky Way, is small but distinguished by Alpha Aquilae (Altair), a first magnitude white star 16 light-years away. The two stars flanking it are Beta and Gamma. Being in the Milky

TWO OF THE NEBULA which cluster thickly between us and the centre of the Milky Way Galaxy, as seen by the 5 m Mount Palomar telescope. Right is the Trifid Nebula (NGC 6514 or Messier 20) and, overleaf, the Lagoon Nebula (NGC 6523 or Messier 8). Both are in the constellation Sagittarius. Nebulae such as these, the Orion nebula and many others, are characteristic objects of the spiral arms of the galaxy where gas and dust has been concentrated by the rotation of the galaxy. The nebulae are the birth-places of the stars, which condense out of their gas and dust. At first the nebulae are dark, but as stars form and begin to shine, the gas and dust also begins to glow. The nebulae are often many light-years in extent and the gas is extremely tenuous —by laboratory standards it would be a very good vacuum.

(Hale Observatories)

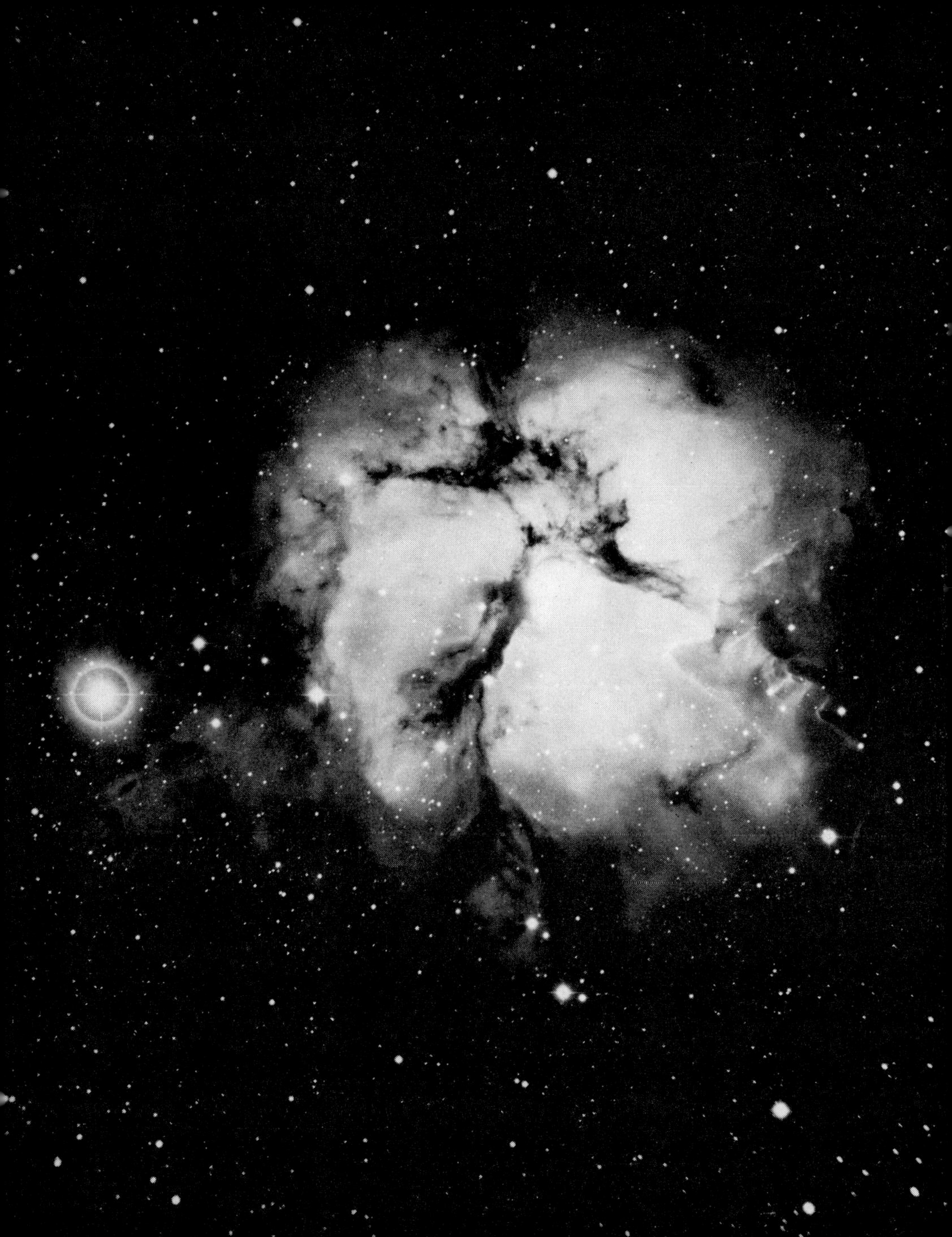

Way, the constellation abounds in doubles, clusters and patches of nebulosity.

North of Aquila is Delphinus, the seven brightest stars fitting neatly into the 7×50 binocular field. Alpha Delphini is an easy double, and a telescope will make a striking double of Gamma.

Capricornus is rather like Ophiuchus—big, but deficient in bright stars. Alpha Capricorni is clearly double to the unaided eye. Four fainter stars are close by. Beta is also a double with each of the components in turn a close double. The part of the constellation toward Piscis Austrinus is a good binocular field.

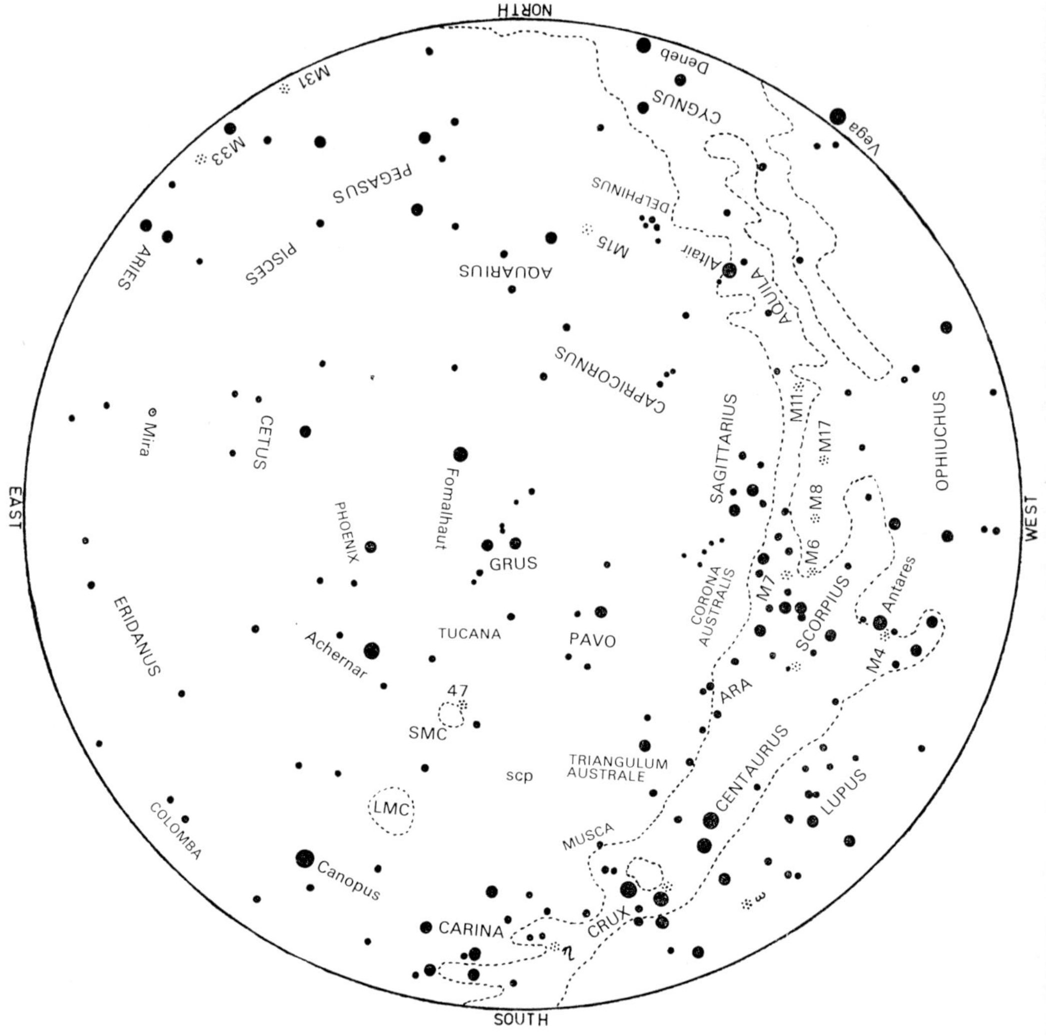

The Sky from New Zealand: **October 1 at 10 p.m.** **November 1 at 7 p.m.**
October 15 at 9 p.m. **November 15 at 6 p.m.**
October 30 at 8 p.m. **November 30 at 5 p.m.**

THE SKY ABOVE US: OCTOBER-NOVEMBER

Sagittarius and the whole Milky Way are now well down to the west, Centaurus, Crux and Carina are low to the south and Canopus is starting to climb east of the pole. The Magellanic Clouds are now a little higher. Eridanus is rising in the south-east and Cetus is due east with Aries, Pisces and Pegasus stretching round to the north and Cygnus due north. Aquarius is high to the north and Grus has reached the zenith.

Aquarius is another big constellation lacking in bright stars—indeed only three are as bright as magnitude three. There are three clusters in a binocular field in the part toward Capricornus. Pavo and Grus are worth getting to know by the distinctive shape of the groupings of the brighter stars, but neither group contains much of interest to the unaided eye or small instrument.

The bright star near the zenith marked on the chart as Fomalhaut (the first syllable is 'fom' and the final 't' should be sounded; it's an Arabic name, not French) is Alpha Pisci Austrini, although the constellation is not marked. Beta and a striking pair of almost equal magnitude can be brought into the same binocular field as Alpha, and moving toward Capricornus another distinctive group can be picked up.

Southern Pole Star

Earlier in this book finding the South Celestial Pole, the point around which the sky rotates, was made a practical exercise. Now let's see if we can find the star that comes closest to marking it, although we have nothing to even approach the magnificence of the northern pole-marker, Polaris. (We can't do anything about this overnight, but the precession of the equinoxes will take care of it—in time. The pole takes 26,000 years to complete one precessional cycle.) A binocular field will show a group of five fifth-magnitude stars close to the pole, four arranged in a trapezium and the fifth a short distance away from one end of the longest side. The short side of the trapezium from this corner ends at Sigma Octantis, which is our apology for a visual pole marker. Sigma should be visible to the unaided eye in good conditions. It makes a wide pair with a seventh-magnitude star which is actually closer to the pole. If the correct pair has been found, Sigma will rotate around its companion as the night proceeds.

While talking about the pole is a good time to become familiar with the term 'circumpolar' and what it means. You will have noticed that a number of objects have never disappeared from the charts, although they were pretty low to the south at some time during the year. Canopus is a good example. Trace its movement on the charts for the whole year and, low though it gets in August-September, it's always there. Any object that does this is termed 'circumpolar'. It means the object never sets; it is always above the horizon.

This comes as a shock to some people, upsetting a fundamental belief that everything rises in the east, crosses the sky and sets in the west. A little thought will show that there must be an exception to this. If the distance of an object from the pole is less than the pole's altitude above the horizon, then obviously that object will never reach the horizon as it sinks on the western side of the pole. It will be at its lowest, but still above the horizon, when directly south, then it will start up again on the east side of the pole.

The altitude of the pole varies with the observer's position. In fact, the altitude is equal to the observer's latitude. All along a given line of latitude, right around the world, the same stars are circumpolar. But if you move north or south, the circumpolar stars change. Only at the equator are no stars circumpolar. There, the poles themselves lie on the horizons. At one or other of the poles only half the celestial sphere is seen, but all stars are circumpolar all the time. The area of the sky which is circumpolar is equalled by the unseen area around the other pole. Please note, though, that when it is said that a circumpolar object never sets, a perfect—that is, a sea-level—horizon is assumed. A circumpolar object may, of course, disappear behind the local geography. That doesn't count.

Incidentally, a star as bright as Canopus begins to do interesting things when it is very low and shining through dense, turbulent air. All sorts of refraction effects take place and it changes colour all the way from red to blue and back again. It sways backwards and forwards as its light passes through wavering currents of air.

It can look for all the world like a flying saucer—and Canopus is so identified with monotonous regularity. The turbulent air above a corrugated iron roof as the heat leaves it after sunset can do strange things to stars quite high in the sky if the observer's line of sight is right. These are effects worth watching for and understanding, both for their own interest and to strangle at birth some of the most irritating of the flying saucer reports. If you become an amateur astronomer you will discover that the more you make a systematic study of the sky, the fewer flying saucers you will see.

Later in the night, when Orion is just rising, Eridanus can be traced from Achernar to near Rigel. There are several interesting star fields along its length and it can be noted how truly like a river this long, meandering constellation is.

Andromeda Nebula

It is impossible to read any reasonably full text on astronomy without finding references to M31, the Great Nebula in Andromeda. Without going into too many details, it's a spiral galaxy comparable with the Milky Way Galaxy and about 2 million light-years away. It is the pride of the northern sky, visible to the unaided eye and certainly the most distant object the eye can see clearly without instrumental aid. It rises just far enough above our northern horizon to be seen from New Zealand. Again, it will need good conditions and a reasonably unobscured northern horizon, but it's marked on the chart and can be found quite easily in relation to some quite unmistakable stars in Pegasus. In Wanganui we can just pick it up in the big telescope, looking almost horizontally across the city's lights, and, frankly, it doesn't look like much, but it is one of the most important objects in the sky and well worth glimpsing, even in poor conditions.

Migrants from the Northern Hemisphere may locate M31 quickly by reference to the four bright stars which form what is known as the Great Square of Pegasus. It is as well to know, however, that while three of these stars are Alpha, Beta and Gamma Pegasi, the fourth is Alpha Andromedae. Beta Pegasi is a bright variable dropping from magnitude 2.1 to 3 irregularly, but usually in something like thirty-five days.

The Small Magellanic Cloud and the cluster numbered 47 are

in the unmarked constellation Tucana. The cluster is the second-best of the globulars, at fifth magnitude fainter than Omega Centauri, but still a naked-eye object and striking in an instrument. Further around the edge of the cloud is another globular, at sixth magnitude right on the border of unaided eye visibility, but strongly condensed and bright in its central regions. These globulars, as mentioned previously, are out in the sphere of the galaxy far from the central plane, but they are part of our galaxy. The Magellanic Clouds are not. They will be discussed in connection with the December-January chart, when they are a little higher in the sky.

The constellation Cetus is now high enough to be viewed in good conditions, so we will look for a famous variable. It is marked on the chart as Mira and shown as a small dot within a circle. There is a proper procedure to be followed here, and it is rather important that it be followed. The reason will be apparent soon. Note the distinctive pattern of the four stars above it and the pair below. Ignore Mira altogether. Now look at this area in the sky, find the six stars noted on the chart and then see if Mira is visible. It may not be, which is why it is important not to try to look for it directly.

Where Algol is the best-known eclipsing variable, Mira is the best-known intrinsic variable—that is, a star which really does vary its light output. Mira was the first such star to be identified. It is irregular both in range and period, but the period is reasonably constant at eleven months from maximum to maximum. The brightness range is highly variable, however, although it is usually visible to the unaided eye for several weeks around maximum when it may be as bright as second magnitude, or as faint as fifth magnitude. It then falls slowly to the ninth magnitude and becomes an object for a quite large telescope. The rise to maximum is faster than the fall to minimum.

GREAT NEBULA IN ANDROMEDA, pride of the northern sky, can be glimpsed low toward the northern horizon from New Zealand. It is a spiral galaxy, similar to our own Milky Way Galaxy. The spiral arms in the densely populated central plane are nearly obscured because we view it almost edge-on. This photograph, from the 1.2 m Schmidt camera at Mount Palomar, shows also the two small galaxies which are satellites of the big spiral. The Andromeda Nebula is also known as Messier 31 and appears on the charts as M31.

(Hale Observatories)

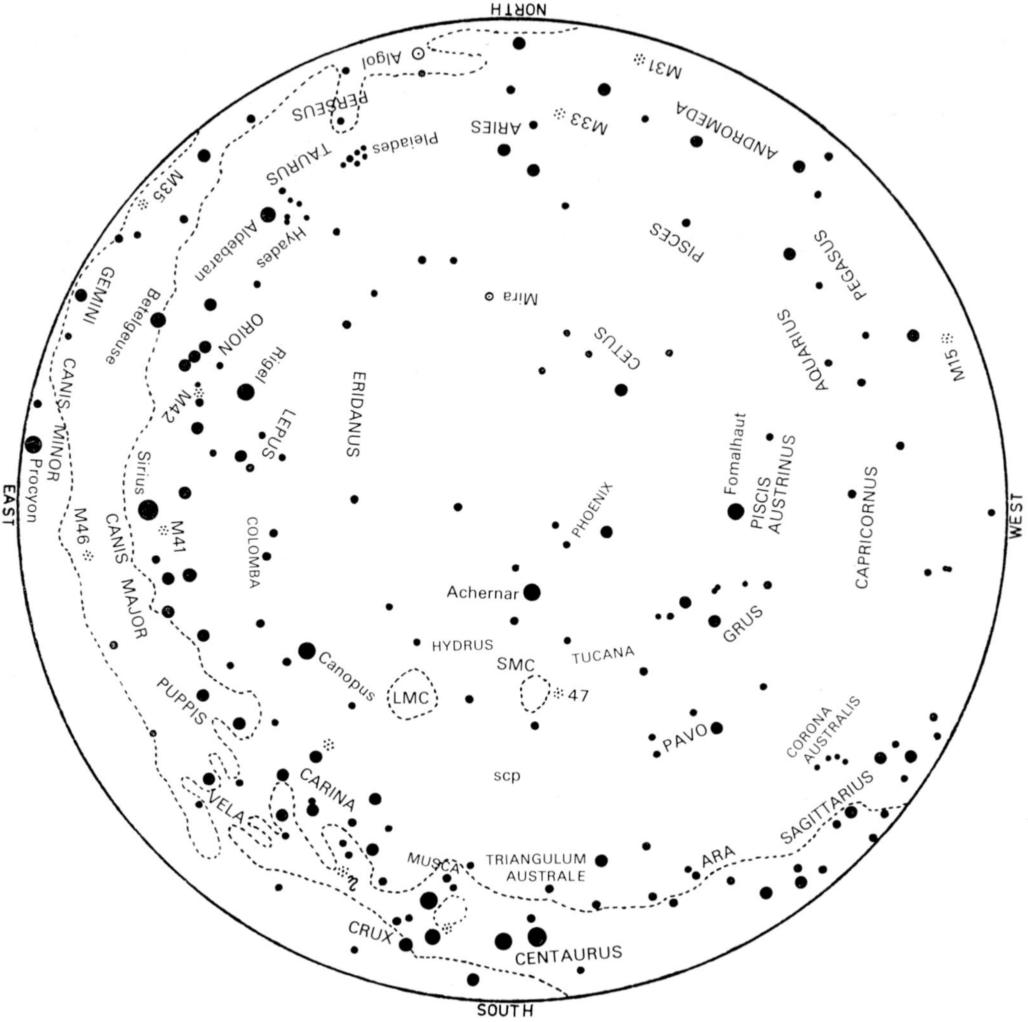

The Sky from New Zealand: December 1 at 10 p.m. · December 15 at 9 p.m. · December 30 at 8 p.m. · January 1 at 7 p.m. · January 15 at 6 p.m. · January 30 at 5 p.m.

THE SKY ABOVE US: DECEMBER-JANUARY

The Milky Way now stretches along the horizon from the north right round the east to the south and south-west. In the south, Carina, Crux and Centaurus are all low but the Magellanic Clouds and Hydrus are high. In the south-west, Sagittarius is setting with Corona Australis, Pavo and Grus above it, and due west, Piscis Austrinus, Capricornus and Aquarius are low. Phoenix and Cetus are only just past the zenith. To the north, Pisces and Andromeda are dropping from sight and Aries is as well placed as we ever see it. In the east a rich region of the Milky Way is rising, bringing in the north and north-east Taurus, Orion, Gemini and Lepus, then on to the south, Canis Minor, Canis Major, Puppis, Vela, Carina and Crux Australis. Eridanus stretches from the east around to the south and Achernar is the only bright star near the zenith.

It was pointed out when discussing Scorpius that this was a particularly rich region of the Milky Way. Now, almost six months later, we look out in the opposite direction at the same time of night, again toward an extremely rich region, the constellation Orion, in a spiral arm of the galaxy.

'Shining Like Stars'

Gabriel Read began one of the most spectacular chapters in New Zealand history on 23 May 1861 when he discovered gold in the gully that has now borne his christian name for more than 100 years. Later he described the first nuggets he saw in his pan as 'shining like the stars in Orion'. It was a singularly apt description. This spectacular constellation lies half in the northern sky and half in the south, the celestial equator passing almost exactly through one of the stars in the belt.

A word about the mythical figure of Orion the Hunter: in the south we see it upside down. But in the north, the red giant Alpha Orionis (Betelgeuse) marks the head, the trunk stretches down to the three stars of almost equal brightness which form the belt, from which hangs the hunter's knife. The blue giant Beta Orionis (Rigel) marks the feet. We tend to ignore the full figure and to see the belt stars, the three fainter objects of the knife and a seventh star as a saucepan-like figure commonly called 'The Pot'.

Almost all the stars we see in this constellation are members of a big, loose association of stars moving through space with a common direction and speed. They, and the host of fainter stars between them which we cannot see, have condensed out of a cloud of gas and dust, the remnants of which still envelop them. Betelgeuse alone of the prominent stars does not belong to the Orion association. At 650 light-years distant, it is 200 light-years closer than the Orion association. Rigel is the sixth-brightest star in the sky and Betelgeuse, although slightly variable, the fourteenth.

The nebula has several centres of concentration, the main one around the middle star of the knife, or the handle of the pot. This is Theta Orionis (M42 on the chart) which is perceptibly hazy to the unaided eye and clearly so in binoculars. In a small telescope the star breaks into four stars of almost equal brightness, known as the Trapezium. This is the brightest part of the nebula and on a clear, dark, moonless night, and with a large telescope, it can be traced over a considerable area. Photographs show that it extends to the boundaries of the constellation and beyond.

Where Stars Are Born

Nebulae are the birthplaces of the stars, the primordial material of the galaxy from which the stars condense. At first a proto-star is huge and diffuse. We would not think of it as a star, but it is stable and self-contained, and shrinking under its gravity to more

GREAT NEBULA IN ORION, or Messier 42 (M42 on the charts), is the object that appears hazy even to the unaided eye in the handle of Orion's sword. This photograph by the 2.5 m telescope on Mount Wilson shows both glowing and dark gas. Hot, blue stars which radiate high levels of ultraviolet light cause the gas in their vicinity to fluoresce. Dust in the nebula shines by reflected light. The bright star at the heart of the nebula (completely obscured in this photograph) is Theta Orionis. In even a small telescope it resolves into a group of four known as the Trapezium. *(Hale Observatories)*

star-like form. It is cold and dark. Slowly it heats up from the energy of collapse and becomes a faintly glowing volume of gas, millions of kilometres in diameter. For aeons the star remains in this condition, perhaps pulsing slowly in and out, but growing always smaller and with the temperature at its core climbing steadily toward a critical limit. When this critical point is reached, thermo-nuclear reactions begin in the hydrogen and the star becomes a giant nuclear reactor. It expands and becomes very hot. If it is big enough, it will be hot enough to radiate most of its energy in the blue part of the spectrum, as Rigel and the Belt stars do; if it is not so big, temperatures will not rise quite so high and it will be a different colour, white like our Sun, or green, or yellow.

These new stars, especially the blue ones, affect the nebula material that remains around them. The dust shines by reflected light and the ultraviolet light causes the gas to fluoresce. The nebula changes from dark, like the Coal Sack, to bright, like this one in Orion or the one around Eta Carinae. In the Orion nebula it is possible to see dark areas where the gas is too distant from the stars to be affected.

Any sort of instrument reveals faint stars in great profusion in this area, as is also the case in Scorpius. This is typical of what is happening all along the spiral arms of the galaxy. The arms are formed by the concentration of gas and dust into lanes by the rotation of the galaxy and in these concentrations the condensation of stars is proceeding rapidly, producing the large numbers of young, hot, blue stars which make the arms of any spiral galaxy stand out so prominently.

Other concentrations of nebula lie around the belt stars, including the much-photographed Horsehead Nebula, but this requires a big instrument for even a glimpse.

The belt star below the handle of the pot is Zeta Orionis. A short distance from it toward the handle is Sigma Orionis, visible to the unaided eye, which a telescope will resolve into six separate stars in two groups of three. The middle belt star is Epsilon, with a group of fainter stars close by, and the third belt star is Delta Orionis, almost exactly on the Celestial Equator. As with the other rich areas of the sky, Orion precludes detailed description with only a simple chart for guidance, but it amply repays careful study. It is one of the best-known constellations in the sky.

Two Notable Clusters

Taking a line from the belt stars across the constellation Taurus two great open clusters can be found, both in the spiral arm. The first and most open of the two is the Hyades Cluster, clearly marked by the young red giant Alpha Tauri (Aldebaran), twelfth brightest star. The cluster looks like an 'A' lying on its side.

Beyond the Hyades but still in Taurus is the Pleiades Cluster which has a curiously unreal or shimmery appearance. This is caused by the combined light of the numerous fainter stars between the bright ones we can see. And how many can we see? That is a question to open up quite a discussion. This cluster is the Seven Sisters of northern mythology and legend has it that only six are now visible. In fact, how many can be seen depends entirely on conditions. There are eleven stars brighter than sixth magnitude (the brightest is fractionally under third magnitude) all of which should be visible to the unaided eye in good conditions, and a further dozen shade gradually down to seventh magnitude, and another seven stars are brighter than magnitude 7.6. This cluster is magnificent in 7×50 binoculars while no ordinary telescope shows it well; a telescope's field is too narrow. Traces of the nebula from which these stars condensed are still present and may be glimpsed in a moderate telescope.

The Maori New Year

Some time in the month we now call May the ancient Maori tohungas, the priestly adepts of the tribes, began a late night vigil, accompanied no doubt by fasting, prayer and ritual, and as dawn approached they anxiously scanned the eastern sky, seeking a glimpse of that strangely shimmering patch of light they called Matariki and we call the Pleiades. If the nights were clear and it was seen early in the month, it was a good omen for the coming year, but if clouds intervened and delayed the sighting it boded ill, for it was from this pre-dawn (heliacal) rising that many of the tribes dated their new year.

Maori culture was not as uniform as it is now sometimes thought to have been and practices differed widely from tribe to tribe. Some dated the year from the sighting of Matariki itself, others from the first new moon after sighting the cluster. Still other tribes

PLEIADES CLUSTER in the constellation Taurus is one of the most beautiful objects in the summer sky. It is clearly visible to the unaided eye as a patch of light with a distinctive shimmering appearance. Almost the whole cluster can be seen in binoculars, which show it to perfection. Telescopes, with their narrow field, can show only a small part of the cluster at a time, as does this photograph of the central region by the 2.5 m telescope on Mount Wilson (compare this with the cover photograph). Traces of the nebula from which the cluster stars condensed are plainly visible. *(Hale Observatories)*

counted their year from the rising of Rigel in Orion at sunset (cosmic rising) which gave them a quite different year from those who followed Matariki. Even among these latter tribes, however, Matariki held a special place in their sky legends, as it has for peoples everywhere, at all times. It is one of the most beautiful groups in the sky.

Supernova Remnant

On a line joining Aldebaran to M35 on the chart, just where it passes a star in the Milky Way, is a hazy patch of magnitude 8.4, known as the Crab Nebula from a rather far-fetched resemblance to a crab, which needs a big telescope to detect. This insignificant-appearing object is the remnant of one of the most spectacular phenomena the universe affords on a time-scale sufficiently short for human comprehension. Perhaps two or three times in a hundred years in a collection of stars as big as the Milky Way Galaxy a hot, giant star becomes unstable and flings a large part of its mass into space in an immense explosion. Within three days, the exploding star, called a supernova, becomes 100,000,000 times more brilliant than the Sun—as bright, in fact, as a whole galaxy. Most of this rise in brilliance actually takes place within a few hours. It holds its peak brilliance for a few days at most, then begins to fade again, and within a few months is only a super-bright star, although it will fade for years toward its final state.

At least three supernovae have been observed in the Milky Way Galaxy in the past 1000 years (and the Star of Bethlehem may have been one too). Others have probably taken place in regions hidden from observation. Most of what is known about these outbursts has been discovered by studying those that have appeared in other galaxies.

Pulsar

Two of the three in our galaxy, recorded in 1054, 1572 and 1604 as brilliant 'new' stars, were too distant to leave visible traces, but the 1054 outburst did: the Crab Nebula. This bright 'new' star was recorded by Chinese astronomers, but apparently passed without notice in the Europe of the Middle Ages. The nebula we see now is an expanding cloud of debris from the explosion of the star which is still visible at its core. This is a white dwarf star, pro-

CRAB NEBULA, photographed in red light by the 5 m Mount Palomar telescope. This object, sometimes referred to as Messier 1 (see description on page 127), is the expanding remnant of the supernova of 1054. The filamentary structure is related to the lines of magnetic force in the gas cloud. The star at the heart of the nebula is a rapidly spinning white dwarf, all that is left of the star that became the supernova and blasted a substantial portion of its mass into space to form the nebula. The interaction of the magnetic field of the star and the gas cloud produces the pulsed radio signal which led to sources such as these being named 'pulsars'. The nebula is expanding at almost 1000 km/sec, but would show no discernible change in a lifetime's viewing in a small instrument. *(Hale Observatories)*

duced, presumably, by the tremendous compression effect of the explosion on the core of the star. The star, spinning rapidly in a strong magnetic field, produces powerful radio waves which reach us in short, rhythmic pulses, from which is derived the name of such objects: pulsars.

Earlier in the year, we were able to glimpse the big, relatively close galaxy M31 in Andromeda. We have seen that stars are sometimes single, sometimes in clusters. So, too, are galaxies—on a vastly greater scale. The Milky Way Galaxy belongs to a cluster called the 'local group'. We cannot see the cluster very well because we are inside it and much of it is obscured by the dense parts of the Milky Way, but M31 is a member, together with two other fairly big ones and perhaps a dozen smaller ones. The smallest of these are so small that they might fairly be called superstar clusters, except that they are self-contained objects, clearly separated from any other.

Two of the intermediate-sized galaxies are quite close to us, however. They occupy extended areas of our sky and can be seen quite clearly with the unaided eye, even across so great a distance. It is these two objects which we are now going to look at.

The Magellanic Clouds

The Magellanic Clouds are galaxies in their own right, and to this extent they are comparable with the Milky Way Galaxy. There are, however, major differences between the two of them, and between each and the Milky Way Galaxy. To begin with, both are smaller, perhaps one-twentieth the mass of the Milky Way. They are close to each other in space (on the galactic scale) and close to the Milky Way Galaxy (160,000 light-years distant), of which they are actually satellites, bound to it by gravity and in immensely big orbits around it. The Large Cloud is basically similar to the Milky Way. It has spiral arms, although only two, and they emerge not from a central nucleus, but from each end of a long 'bar'. As we see the cloud edge-on, this 'barred spiral' structure is not apparent. The Small Cloud is an 'irregular' galaxy, having no internal structure at all, but is instead a giant cloud of stars unevenly distributed through a volume of space which is not even approximately spherical.

THE MAGELLANIC CLOUDS, satellite galaxies of the Milky Way Galaxy and two of the most important and striking objects in the southern sky. Long used for navigation in southern seas, they form an almost equilateral triangle with the South Celestial Pole. The Large Cloud (above) is a barred spiral, two arms emerging from either end of a long bar. The bar is readily apparent in the photograph, but the arms are obscured by the fact that we see the galaxy almost edge-on. The Small Cloud (right) lacks a distinctive structure. Both clouds contain open and globular clusters, bright and dark nebulae and all the objects familiar to us from the Milky Way. The Large Cloud is rather more spectacular to observe than the Small Cloud. To the right of the Small Cloud

on the plate is the globular cluster 47 Tucanae (NGC 104), second-best of the southern globulars, and above the cloud is another globular, NGC 362. At magnitude six, it is a full magnitude fainter than 47 Tucanae. These globular clusters are objects in the Milky Way Galaxy which happen to lie in much the same line of sight as the Small Cloud and thus appear close to it in the sky. The Clouds' own globular clusters are too faint at such a distance to be detectable in anything but the largest instruments. Both plates were taken by the 20.3 cm Schmidt camera at Mount Stromlo Observatory near Canberra in Australia. *(Mt Stromlo and Siding Spring Observatory)*

In other ways both clouds do resemble the Milky Way Galaxy. They contain the same sort of objects and we can identify open clusters of the Pleiades and Hyades type, globular clusters such as Omega Centauri and dark and glowing nebulae like the Coal Sack and Great Nebula in Orion. In all these things the Large Cloud can be regarded as a miniature of our galaxy with a slightly different form, but the irregular Small Cloud is a basically different type of galaxy.

Any star that we can see as a point of light when we look at the clouds is in the Milky Way Galaxy, between us and the Magellanic Clouds, but there is one object of such breath-taking splendour that it is quite clearly visible to the unaided eye across 160,000 light-years. Near the eastern (left hand) edge of the Large Cloud (as seen at this time of year) is a bright, perceptibly hazy spot of light, an Orion-type nebula. In fact, it appears to us almost as bright to the unaided eye as the Orion nebula, but think about the relative distances for a moment: Orion is 850 light-years away; the Large Magellanic Cloud, 160,000 light-years distant. Obviously, the nebula in the cloud is many times brighter. It is so bright that, were it only 850 light-years away, it would be a glowing cloud in our sky as big as the whole constellation of Orion, and its light would be enough to work outside and to read by whenever it was in the night sky—if we could speak of 'night' in such circumstances.

The nebula is glowing gas, but it derives its light from a star or stars within it. Even a hot, young, blue giant star such as Rigel doesn't pour out enough energy to be visible across so great a gulf, and neither does so powerful a radiation source as the mysterious object at the heart of the Eta Carinae Nebula. The stars that power this nebula are a cluster of one hundred or more supergiants, spread through an irregular volume of space about 200 light-years across. It is several hundred times brighter than the great globular cluster Omega Centauri and its gravitation is such that it could almost be said to rank as its nucleus. It is, however, at one edge of the cloud and barred spiral galaxies have no central condensation. In a telescope of reasonable size the nebula shows a branched, looped structure after which it has been named variously Loop Nebula and Tarantula Nebula. The supergiant cluster is 30 Doradus and the nebula NGC 2070.

Tracking straight across the diameter of the cloud for 3.5 degrees (Alpha to Beta Centauri is 4.5 degrees) brings another cluster into view, NGC 1910, also containing one hundred or more giants and supergiants. It is not, however, quite so splendid an object as 30 Doradus, its supergiants being fewer. One, however, tries very hard to make good that deficiency. It is nothing less than the brightest star yet discovered in the universe, pouring out 1,000,000 times the light of the Sun. Its designation, S Doradus, tells us at once that it is variable (all Roman capitals beyond Q designate variable stars) but as it ranges only between magnitudes 8.2 and 9.4 it is always accesible to small instruments. It may also be a double star.

Although immensely more powerful than Eta Carinae, S Doradus has this in common with it: it is not known by what mechanism it sustains its immense energy output. The only comparable objects known are supernovae, but they fade very quickly. S Doradus goes on and on, yet it is pouring out energy at a rate that suggests that it could not possibly sustain it for more than 100,000 years. Within this period, only an instant on the cosmic time-scale, it must surely either burn out or become unstable. It is impossible to imagine what this star, now burning with the power of a supernova, might then do.

It is interesting to compare this estimated lifetime of 100,000 years with the 160,000 years that its light takes to reach us. The star has almost certainly ceased to exist in its present apparent form and the light which carries the message of its final fading or catastrophic death is probably already on its way. It could arrive tonight—or not for 100,000 years.

As might be expected, the Magellanic Clouds are much bigger than they appear; only the brightest central regions are visible to the unaided eye. Photographs show the diameters to be about three times what we see, and they are separated from each other by only about the diameter of the large cloud.

In November 1975 the discovery of a new member of the local galaxy cluster was reported, a mini-galaxy only a fraction of the mass of the Milky Way Galaxy and so faint that it was detected only by a study of hydrogen gas clouds. It is believed to be significantly closer than the Magellanic Clouds, but does not rival them in importance.

CONCLUSION

The purpose of this book was to introduce you to astronomy and to the southern sky in particular. If you have gone outside at night, searched for and found some of the objects described, then it has succeeded.

If you have become sufficiently interested to read further and to extend your knowledge, you will find there are many rewards. The literature is extensive and will tell you all you are likely to want to know. The bookshelf, however, can be a trap. Books can be informative, fascinating, absorbing—but they are not astronomy.

Some years ago I examined a boy scout for an astronomy badge. I awarded it to him with honours; it was impossible to do anything else. He knew the distance, mass, density, magnitude (both apparent and absolute), position and any other relevant characteristic of everything I asked him about. He knew his books backwards, but I don't think he had ever seen the sky at night.

Books are among the most marvellous things ever produced. They contain all the questions men have ever asked, and all the answers that have yet been found, but they are not astronomy. Astronomy is stars and the stars are in the sky.

I hope I have led you there.

Appendix One

FINDING THE PLANETS

For the reasons explained more fully in the text, the reader requires current data to permit him to identify the planets among the stars. There are two main sources of this information in a form both suitable for the beginner and compatible with this book:

1. *The Astronomical Handbook* issued annually in December or January by Carter Observatory, Wellington, and available direct from the observatory, through astronomical societies or from selected bookstores. It costs less than $1.00.
2. The *Newsletters* of the Royal Astronomical Society of New Zealand and most district societies.

Purchasers of the *Handbook* will find six charts which alternate with the six in this book to form a complete set of twelve, one for each month of the year, thereby reducing the corrections that need be applied for the passage of days. The planets are marked on the *Handbook's* charts.

The *Newsletters* of the Royal Astronomical Society of New Zealand, received by all members, do not contain charts but do provide considerable data additional to that in the Carter Observatory *Astronomical Handbook*.

The newsletters produced by the various district societies do not usually contain charts either but do provide sufficient information which, in conjunction with the charts in this book, will enable you to identify the planets.

The *Astronomical Handbook* has the advantage that all information is in convenient book form. Articles on a variety of astronomical subjects and data on objects other than the planets are included.

The newsletters, often issued monthly, have the advantage of being able to provide current data on new discoveries, such as comets. There are also a number of other advantages to be derived from society membership (see Appendix Eight).

Appendix Two

NEW ZEALAND METEORS AND METEOR RADIANTS

Seven authenticated meteorites have been found in New Zealand, of which only one was seen to fall. In addition, a small object found at Avondale, Auckland, in October 1956, after an event which could have been the fall of a small meteor, was shortly after confirmed as a meteorite. In 1974 a published account of New Zealand meteorites included this object and the circumstances under which it was found. That year the author initiated an investigation which resulted in this object being reexamined and determined to be definitely not meteoritic. This does not entirely preclude that the Avondale event was a meteor strike, but if so, no fragment of the meteor was found. The investigation also determined that two objects held in New Zealand museums and, although never properly examined, believed to be meteorites found locally, were definitely of terrestrial origin.

It is now unlikely that any unconfirmed meteorites exist in institutions, but because one object only recently authenticated had been held in private hands for many years, it is at least possible that other New Zealand meteorites are held privately.

Because meteorites are of some considerable importance to science, anyone who is in possession of an object believed to be a meteorite is invited to submit it for testing and authentication by the New Zealand Geological Survey. Testing can be non-destructive if the owner wishes, although it would be preferable if permission was given for more intensive testing of a fragment from the specimen. After testing, the specimen would be returned, with a report of the finding. Preliminary inquiries may be made either direct to the Geological Survey, or to the author.

The seven New Zealand meteorites currently known to exist are:

WAIRARAPA. A 5.8 kg stone found in 1863 by a farmer, W. H. Donald, just south of Masterton. The greater part of the specimen is in Auckland Museum, with fragments in a number of other New Zealand institutions and also overseas.

MAKAREWA. A 2.3 kg stone found at Makarewa Junction near Invercargill in 1879 by the brothers A. and I. Marshall while working on the railway line. Has been broken into numerous fragments and distributed among New Zealand institutions, and probably overseas.

MOKOIA. Originally two large fragments of a total weight of 4.6 kg, the larger heavier by 28.35 g, and an unknown number of smaller ones. Stone, and of the rare carbonaceous chondrite type of which only thirty-six exist (twenty-three of these were observed to fall).

This is the only New Zealand meteorite seen to fall. The event occurred at Mokoia, 10 km south of Hawera, South Taranaki, at 12.30 p.m. on November 26, 1908. The first warning was a thunderous explosion, reportedly heard from central Taranaki to the mouth of the Rangitikei River and inland to the villages on the Wanganui River, followed by a series of sharp, lesser explosions 'like rifle shots'. A brilliant streak of white light was seen to disappear to the south and a number of fragments fell to earth. People on Castlecliff Beach, Wanganui, saw the main mass of the meteor plunge into the sea about half a mile off shore.

Of the two big fragments, the larger is intact and on display in Wanganui Regional Museum; the smaller has been completely broken up and widely distributed in New Zealand and overseas. A smaller, intact fragment is in Taranaki Museum, New Plymouth. The author has been unable to discover if other fragments exist, but would be very interested to hear of any. This type of meteor is rare and important for the amount of carbon contained in the form of quite complex molecules believed to have a bearing on the origin of life on Earth.

MORVEN. A 7 kg stone ploughed up at Morven, South Canterbury, by William Stewart, a farmer, in 1925. The largest part is in Otago Museum, with smaller fragments in other institutions.

VIEW HILL. A 33.6 kg nickel-iron type found at View Hill near Oxford, Canterbury, in 1953 by C. C. Anderson, a farmer. Presented to Canterbury Museum in 1954. Fragments are in several New Zealand institutions.

WAINGAROMIA. A 9.2 kg nickel-iron meteorite found at Waingaromia, 23 km north-west of Tolaga Bay, by A. Hedley Reeves, now of Gisborne, in the early years of this century. Not authenticated until presented to Canterbury Museum in 1970. The greater part is still in Canterbury Museum, and a small part at the Smithsonian Institution in the United States.

DUNGANVILLE. 55 kg nickel-iron, the most massive discovered to date. It was found early in 1976 by Mr Ted Dowie, of Greymouth, when prospecting for gold in the Dunganville area, south-east of Greymouth. He found a small but extremely heavy rock in the riverbed and chipped off a sample which he sent to the Geological Survey. It was quickly identified as meteoritic and the object was recovered and despatched to the survey laboratories. After thorough examination it will go to the Canterbury Museum. There were early indications that it had unusual characteristics.

Southern Meteor Radiants

In this table the radiant is on the celestial equator or south of it, although the constellation may be mostly north. Some are in the morning sky at the time of the shower. The name of the shower is derived from the constellation in which the radiant appears. Where two or more radiants occur in one constellation, they are identified by the designation of the nearest star or the month of appearance.

The date of maximum is uncertain in all cases. All far southern showers are badly observed, and observational evidence of southern radiants not listed would be valuable.

Name of Shower	Date of Maximum	Normal Limits	Comment
Corona Australids	March 16	March 14-18	
Eta Aquarids	May 5	May 1-8	
Ophiuchids	June 20	June 17-26	
Capricornids	July 25-26	July 10-August 15	Bright meteors
Delta Aquarids	July 27-28	July 15-August 15	Radiant is double
Pisces Australids	July 31	July 15-August 20	
Alpha Capricornids	August 2	July 15-August 25	Meteors yellow, bright
Iota Aquarids	August 6	July 15-August 25	Double radiant
Phoenicids	December 4-5	December 4-5	

Northern Radiants

These radiants are north of the celestial equator but observable from the south, otherwise as above.

Name of Shower	Date of Maximum	Normal Limits	Comment
April Lyrids	April 22	April 19-24	Bright meteors
June Lyrids	June 16	June 10-21	Blue meteors
Orionids	October 21	October 16-26	Multiple radiant; leave persistent trails
Taurids	November 8	October 20-November 30	Double radiant; many bright meteors
Leonids	November 18	November 15-19	
Geminids	December 14	December 7-15	

All showers are unspectacular when the altitude of the radiant is low; some northern radiants are never high enough to show to advantage. Shower characteristics are highly variable from year to year. The data sources listed in Appendix One will indicate when conditions are favourable and a good display may be expected. See also Appendix Five (Photography).

Appendix Three

SOLAR AND LUNAR ECLIPSES AND INNER PLANET TRANSITS TO THE YEAR 2000

Solar Eclipses: Only two solar eclipses are visible from New Zealand in the remaining years of this century, a partial eclipse on 4 February 1981, and an annular eclipse on 16 January 1991.

Inner Planet Transits: Both Mercury and Venus on occasions pass directly between Sun and Earth and are visible as black dots against the solar disc.

Edmund Halley, famous for far more than his comet prediction, devised a means of determining the (then unknown) Earth-Sun distance by observations of Venus transits. These transits, however, take place only at intervals longer than a human lifetime, and none occurred at a time when Halley could make the required observations. He so imbued the infant Royal Society in London with the importance of the work that the society determined that the transit of 1769 should be carefully observed.

The method depends upon the differences between observations obtained from widely separated points and it was to obtain one such set of observations at Tahiti that the barque *Endeavour* sailed from England under the command of Lieutenant James Cook, R.N. This was the prime objective of the voyage, and it was only after these observations had been made that Cook was to sail on into the South Pacific in search of the lands discovered earlier by Abel Tasman.

The observations at almost all stations were successful and the radius of Earth's orbit was determined within about 5 percent, a figure which would have been considerably improved if the observations had not been badly handled by the calculators.

The Venus transit observations are commemorated by the name Venus Point at Tahiti, and observations of a later transit of Mercury (on 9 November 1769) by Cook's first expedition provided the name of Mercury Bay, near Whitianga. But for the astronomical impetus, the rediscovery of New Zealand may have been delayed for many years—and may not have been made by a British expedition.

The next Venus transit is in June 2004.

Mercury transits occur with much greater frequency. There are three in the remaining years of this century: 13 November 1986, 6 November 1993, 15 November 1999. Not all may be observable, at least in their entirety, from New Zealand.

Lunar Eclipses: The following lunar eclipses will be visible from New Zealand. In some cases the Moon will rise and in others set in eclipse. A partial eclipse is one in which the Moon is never entirely immersed in the umbral shadow. All eclipses are total unless otherwise stated. The figure in parentheses indicates the length of total phase in minutes; date given is day of mid-eclipse.

24 March 1978 (90)	20 February 1989 (76)
6 September 1979 (52)	9 February 1990 (46)
17 July 1981 (partial)	6 August 1990 (partial)
9 January 1982 (84)	21 December 1991 (partial)
6 July 1982 (102)	4 June 1993 (98)
30 December 1982 (66)	29 November 1993 (50)
25 June 1983 (partial)	15 April 1995 (partial)
28 October 1985 (42)	16 September 1997 (66)
24 April 1986 (68)	28 July 1999 (partial)
17 October 1986 (74)	21 January 2000 (84)
27 August 1988 (partial)	16 July 2000 (102)

The diameter of the umbral shadow is much greater than the diameter of the Moon. The length of totality, therefore, depends upon whether the Moon passes centrally through the shadow, giving the longest period of totality, or close to an edge, with a shorter time fully immersed.

Precise data for inner planet transits and eclipses in any year will be available in the publications listed in Appendix One.

Appendix Four

MAKING A TELESCOPE

It is possible for a person who is reasonably handy and who is prepared to work to close tolerances to grind a mirror for a reflecting telescope, build it into a tube and mount the completed instrument. Telescopes up to 15 cm (6in.) may be constructed without exceeding the facilities available in a moderately well-equipped home workshop. It will be necessary to have a few jobs done by tradesmen and specialists (i.e. aluminising of the completed mirror to produce a highly reflective surface).

For sizes above 15 cm, the magnitude of the task involved in building the telescope and mounting increases rapidly and exceeds what a home workshop can cope with. If the telescope maker does not have access to a fully-equipped engineering workshop this work will have to be paid for, and may make the project prohibitive.

An extensive literature exists on the subject, to which the best access is through membership in astronomical societies and their libraries. Most societies include members who have had experience in mirror grinding and telescope construction.

It is fair comment that a number of books grossly oversimplify the problems involved and lead to the production of badly corrected mirrors, which in turn lead only to frustration and disappointment in the performance of the completed telescope. This oversimplification occurs most frequently in books whose theme is 'make yourself a telescope' rather than more general instruction in the optical and mechanical techniques involved. The books given in the bibliography do not exhaust the list of reliable publications, but they at least can be used with confidence (the 'Amateur Telescope Making' series is authoritative). Although enthusiasts have produced good mirrors at a first attempt without assistance, because of the unfamiliar techniques and the amount of subjective judgment involved, it is highly advisable to obtain experienced guidance.

Appendix Five

PHOTOGRAPHY

Much of astronomy today is photographic. It is beyond the scope of this book to become involved in useful photographic work possible for the amateur, although there is increasing scope for such activity. The instrumental requirements are much more severe than for visual work. Those who are seriously interested are referred to Appendix Seven, to the bibliography and to the astronomical societies.

For those with ordinary cameras and the urge to extend their picture-taking to astronomical subjects, several possibilities exist.

A camera with a lens operating at least as fast as f4.5 and preferably f3.5 or faster, with a good lens hood, fast film and mounted on a firm tripod can be used for pictures which, while having no scientific merit, can be both attractive and informative. An exposure as short as 10 seconds on the fastest film will record the brighter stars without showing excessive movement.

Exposures of the order of one hour will produce long trails as the turning Earth sweeps the field of view across the sky. Facing north, these will be almost straight lines, parallel to the celestial equator. Facing south, they will be segments of arcs centred on the south celestial pole. Buildings and trees may make interesting foreground silhouettes. Meteor trails may appear on such long exposures, especially during a meteor shower. If so, the result may be not only of interest but of real value.

In theory, if a telescope is correctly focussed and a camera focussed at infinity is held behind it, the object in the telescope can be photographed. In practice, the difficulties multiply. The camera can be hand held only for the Moon. At full Moon expose as for grey stone in full sunlight, but open the diaphragm much wider at lesser phases. Correct focussing can be very difficult and so can lining-up. A single lens reflex camera overcomes the latter difficulties, but some means of holding the camera rigidly in position is virtually a must. Clamps and various attachments can be bought.

See the Bibliography in Appendix Ten for further reading on this subject.

Appendix Six

RADIO ASTRONOMY

At time of publication there were radio astronomy projects being carried out at Auckland University on the Sun, and at Otago University on Jupiter and other objects.

Radio astronomy is another branch of the science which virtually began with amateurs, certainly in the sense that the pioneers were not professional astronomers, and much useful work has been done by amateur radio observers. This has been largely in the field of observations of the occultation of radio sources such as the Crab Nebula by the Moon, especially when the source is occulted as 'seen' from the amateur's observatory but not from any professional radio observatory.

Like many branches of astronomy, however, the minimum instrumental requirements have increased greatly in recent years. Books published in the Northern Hemisphere may be especially misleading in this field, as they often do not allow for the extensive (and often pioneering) work done in Australia. Radio astronomy is one branch in which the south has established a lead. The effect of this is to reduce still further the scope of the southern amateur.

Useful work aside, there is no reason why the radio-orientated amateur should not explore the sky as fruitfully as his optical counterpart. It is a specialised field, however, and one in which the author has no personal knowledge. An extensive literature exists, readily reached through public or society libraries. Amateur radio magazines appear to have a heavier commitment to radio astronomy than publications devoted mainly to optical astronomy. Some articles on radio astronomy do appear in *Sky and Telescope* magazine.

Appendix Seven

USEFUL WORK FOR THE AMATEUR

Astronomy is unique among the sciences for the contribution made to it by amateur observers. If the era is long ended in which discoveries as diverse as the planet Uranus and the sunspot cycle could be made by amateurs, it is equally true that the keen, well-equipped amateur can still make useful contributions in the modern era of large telescopes and computerised data processing.

The space age effectively ended one major field of amateur activity, lunar and planetary drawing, and there is no longer a call for simple sunspot counts. Indeed, the emphasis has changed so rapidly that almost the only printed material now relevant is recent articles in astronomical society journals and amateur-orientated publications.

The principal fields of amateur activity are now:

Visual observations of variable stars. The 'standing watch' on a large number of variables has been handed over to amateur organisations to free professional time and instruments for more difficult objects. The Variable Star Section of the Royal Astronomical Society of New Zealand is one of the most active and probably the best organised of these organisations. Each observer, using a prescribed technique, makes a periodic check on the variable stars on which he has elected to work. The results are forwarded to the section headquarters and there processed and made available to professional observatories as the basis of their observing programmes. The work is equivalent to that of a highly skilled laboratory technician and requires considerable dedication on the part of the observer.

Visual observation and timing of lunar occultations. The Moon periodically obscures bright stars in its passage across the sky. Careful timing of the disappearance and reappearance of the star, particularly in 'grazing' occultations when the star is barely obscured at the limb of the Moon, yields a variety of geophysical and astronomical data.

Comet hunting and visual observation of comets. It is only in recent years that routine photographic surveys by professional observatories have displaced amateur observers as the principal discoverers of comets. Many amateurs still make systematic searches of the sky, or selected areas of it, in search of comets and their efforts are rewarded by discoveries averaging perhaps one per 300 hours of observing time in the Southern Hemisphere (much longer in the north). Comet hunting is weighted rather heavily in favour of the southern amateur, because fewer professional photographic surveys are being made in the south.

Drawings, brightness estimates and photographs of comets, both new discoveries and known comets making predicted returns, are still required by professionals and can be contributed by amateurs.

Photographic and photoelectric observations. The instrumental requirements are rapidly becoming more sophisticated, but many New Zealand amateurs and societies have installed suitable equipment. Photography is devoted mainly to routine survey work for novae in selected areas of the Milky Way and in the Magellanic Clouds, to searches for previously unrecorded variable stars, comet hunting and monitoring of flare stars. Photoelectric photometers are used to monitor variable stars. Any reader interested in such work should approach a district astronomical society or the appropriate section of the R.A.S.N.Z.

Skills other than purely astronomical can also be put to good use. Those with mathematical ability, access to computers or able and willing to act in a consultant capacity for those building optical, mechanical or electronic equipment could give valuable assistance to New Zealand astronomers, both amateur and professional.

In fact, the amateur contribution to astronomy in New Zealand is now so great that the professional astronomers have dropped the word 'amateur', lest it be misleading, and prefer to use 'part-time astronomer' as being a more accurate description of those colleagues who work at an ordinary job during the day and spend their nights (weather permitting) making astronomical observations.

Appendix Eight

JOIN AN ASTRONOMICAL SOCIETY

Astronomical societies in New Zealand are organisations of people interested in astronomy, or some aspect of it. There is a strong professional element in the Royal Astronomical Society of New Zealand, but the majority of members are amateurs, almost all of them interested in rather than actively engaged in some astronomical activity. District societies to an even greater extent are groups of people whose interest and knowledge is no greater than that evident in any other amateur organisation. Any reader who feels that something has been learnt from this book and who wishes to learn more should join a society, the local society if one exists, or the Royal Astronomical Society of New Zealand. Subscriptions vary, but in 1976 ranged between $10 for senior and $5 for student membership of the R.A.S.N.Z. (reduced to $8 and $4 for prompt payment), and $2.50 senior and $1 junior membership for the Wanganui Astronomical Society.

Almost all societies hold regular meetings, publish a regular newsletter, have library facilities and operate an observatory of some kind. Some have instruments which can be loaned to members. Members wishing to accomplish, or assist with useful work are encouraged and helped. Those with more advanced problems, either astronomical or in related fields such as optics, mechanical engineering, instrumentation and electronics, can reach a wide range of professional skills through district societies and their affiliations with the R.A.S.N.Z.

Astronomical knowledge and activity, both world-wide and in New Zealand, is expanding rapidly. Membership in an astronomical society enables an individual to keep up with what is happening and, perhaps, to make some contribution to it, if in no other way, by becoming a member of that informed body of public opinion without which no science can flourish.

Following is a list of New Zealand societies and their addresses. Some groups are restricted to members or students of the parent organisation, but may serve as a source of information or further contacts if tactfully approached.

Royal Astronomical Society of New Zealand, P.O. Box 3181, Wellington.

District and Local Societies

Ashburton Branch, Canterbury Astronomical Society, C/- 20 Russell Avenue, Ashburton.

Astronomy and Geophysics Section, Wellington Branch of Royal Society, P.O. Box 3085, Wellington.

Astronomy Section, Nelson Branch of Royal Society, P.O. Box 233, Nelson.

Auckland Astronomical Society, P.O. Box 2858, Auckland.

Canterbury Astronomical Society, P.O. Box 25137, Christchurch.

Christ's College Astronomy Club, C/- Christ's College, Christchurch.

Dunedin Astronomical Society, P.O. Box 6019, Dunedin.

Friends of Mount John University Observatory Society, P.O. Box 85, Timaru.

Gisborne Astronomical Society, P.O. Box 710, Gisborne.

Hamilton Astronomical Society, C/- 22 Carlton Street, Te Awamutu.

Hastings Astronomical Society, C/- 908 Tawa Street, Hastings.

Hawke's Bay Astronomical Society, C/- 3 Logan Avenue, Napier.

Marlborough Astronomical Society, C/- 106 Maxwell Road, Blenheim.

National Committee for Student Astronomy, P.O. Box 5817, Auckland.
New Plymouth Astronomical Society, P.O. Box 818, New Plymouth.
New Zealand Interplanetary Society, P.O. Box 77, Christchurch.
North Otago Astronomical Society, C/- D. G. Hamer, Alma, 6D, R.D., Oamaru.
Palmerston North Astronomical Society, P.O. Box 398, Palmerston North.
Palmerston North Boys' High School Astronomical Society, P.O. Box 549, Palmerston North.
Southland Astronomical Society, P.O. Box 48, Invercargill.
Taita College Astronomy Club, C/- 9 Foster Street, Taita.
Tawa College Astronomical and Mathematical Society, Tawa College, Tawa.
Wanganui Astronomical Society, C/- Mr C. W. Venimore, Morgan's Road, R.D. 2, Wanganui.
Wanganui Collegiate School Astronomical Society, C/- Collegiate School, Private Bag, Wanganui.
Wellington Astronomical Society, P.O. Box 4085, Wellington.
Wellington College Observatory, Wellington College, Wellington 1.
Wellington College Old Boys' Astronomical Society, C/- 21 Washington Avenue, Wellington 2.
Wellington Planetarium Society, P.O. Box 1177, Wellington.
Whangarei Astronomical Society, P.O. Box 19, Whangarei.
Whakatane Astronomical Society, C/- 21 Goulstone Road, Whakatane.
Carter Observatory, Botanic Gardens, Kelburn, Wellington (P.O. Box 2909, Wellington 1), conducts public sessions each Tuesday from March through October at 7.30 p.m. The format is a lecture and/or film, with a viewing session at the 23 cm refractor if weather permits. There is no organisation attached to the observatory, but the several astronomical organisations based in Wellington all enjoy a close liaison with the observatory. It should be clearly understood that membership in these organisations confers no special privileges at the observatory.
NOTE: The Royal Society of New Zealand, which has a number of branches throughout the country, some of which have astronomy sections, is not to be confused with the Royal Astronomical Society of New Zealand. The Royal Society is the senior scientific organisation in New Zealand and deals directly with the Government, and its scientific equivalents overseas, where necessary, on behalf of its member bodies and affiliates, of which the R.A.S.N.Z. is one.

Appendix Nine

MAORI ASTRONOMY

There is no doubt that many among the ancient Maori accumulated a vast amount of empirical astronomical knowledge and were familiar with the daily and yearly cycles of the sky to a far greater extent than most modern New Zealanders. It is equally certain, however, that this was esoteric knowledge, generally confined to the tohungas (priestly adepts) of the tribes just as it is to the relative few actively interested in astronomy today.

In Maori mythology the Earth and all its creatures were derived from sky-beings, as were the visible heavenly bodies, the 'Whanau Marama', the 'Light-giving Family' or 'Children of Light'. The creation stories were, of course, purely allegorical.

In this mythical-astrological approach to astronomy, the heavenly bodies appear to have been regarded as worlds and believed to have been inhabited. Even the first part of this cannot of course be regarded as 'scientific', but before smiling at the second part, it is well to remember that so famous an astronomer as William Herschel, late in the eighteenth century, thought most, if not all, heavenly bodies to be inhabited—including the Sun, which the Maori most emphatically did not.

The practical manifestations of Maori astronomy were of a sufficiently high order to form a substantial part of a reliable navigation system which permitted great voyages of exploration and return across the Pacific Ocean. These were among the greatest accomplishments of any supposedly 'primitive' people and the navigation involved was much more complex than has often been realised by people unfamiliar with the problems.

For instance, the sailing instruction for the voyage from Hawaiki to Aotearoa (New Zealand) which a European recorded as a simple bearing on the star Canopus is, in fact, useless unless it is also stated at what time of year and time of night Canopus gave the correct course. It was then one of the navigator's more skilled tasks to select correctly those stars which successively occupied the position of Canopus as the daily and yearly rotation of the sky proceeded.

Unfortunately, little effort was made to record Maori astronomy until after the musket, disease, alcohol, a collapsed economy and European impact generally had all but destroyed Maori culture.

The missionaries for the most part were actively hostile to Maori astronomy because of its astrological and religious associations and even those who did attempt to record Maori culture as they found it (perhaps most notably the Rev. Richard Taylor at Wanganui) were inhibited by both their knowledge and their ignorance. As a result, records comprise little more than lists of names of planets and northern stars—that is, those objects which the inquirer could identify and for which he already knew a name. Almost no attempt was made to understand and record the nature and substance of Maori astronomy and this knowledge must now be regarded as lost.

Almost the entire surviving body of Maori star lore is contained in the National Museum of New Zealand (formerly the Dominion Museum) monograph, *The Astronomical Knowledge of the Maori* by Elsdon Best. It has recently been reissued after being out of print for many years.

Best, a member of the museum staff from 1911 to his death in 1931, made a massive contribution to our knowledge of the ancient Maori, but freely acknowledged that astronomical inquiries were made too late and with insufficient thoroughness. Much of what was recorded is useless or suspect.

Because Maori culture was not uniform from tribe to tribe, it appears that certain names were used for different bodies in different places. Few of the investigators, with the exception of Richard Taylor, show any great knowledge of astronomy, however, and the confusion may be largely, if not entirely, on the European side. Esoteric knowledge such as astronomy is quickly lost when a culture collapses and it is unlikely that significant unrecorded information is accessible today. The possibility should not, however, be entirely dismissed and if the confidence of tribal elders can be won, something of value could yet be salvaged by an inquirer prepared to devote time and effort to the task.

Appendix Ten

BIBLIOGRAPHY

Recommending further reading is a problem because individual backgrounds vary and individual requirements are different. The following selection is not exhaustive and most individual titles are not necessarily superior to others covering similar ground. The books listed, however, are selected to give a reliable guide to readers wishing to extend their knowledge of practical, theoretical and instrumental astronomy.

For anyone who finds astronomy becoming a major interest and who wishes to embark on a serious and extensive study, some titles listed, principally those containing calculations, have textbook status. Extensive reading is recommended. A number of inconsistencies, real and apparent, will be detected and in resolving them a great deal will be thoroughly learned.

Astronomy is a rapidly evolving science and progress in some fields is extremely rapid. Older books should not be rejected out of hand as of little value, but obviously the more recent the book, the more likely it is to represent current thinking. Most books published in recent years have been outdated in at least some sections even before they went on sale.

Observing:

Norton, A. P. *Star Atlas and Reference Handbook.* Gall and Inglis, 16th edition, 1973. Contains star maps and lists of interesting objects, plus a tremendous amount of general astronomical information. Older editions are entirely usable, but the rewritten latest edition is superior in some respects.

Staff of the Carter Observatory. *Astronomical Handbook.* Compiled annually and available for the ensuing year in December-January.

Hartung, E. J. *Astronomical Objects for Southern Telescopes.* Cambridge University Press, 1968. An exhaustive list of objects accessible to small and moderate telescopes in the southern sky.

Warner, L. *Astronomy for the Southern Hemisphere.* A. H. & A. W. Reed, 1975. Good background and theory.

Muirden, J. *Astronomy with Binoculars.* Faber and Faber, 1963. A full treatment of all solar system objects in terms of what binoculars will show for northern constellations, but still has useful data for observers in New Zealand.

Menzel, D. *Field Guide to the Stars and Planets.* Collins, 1962. An exhaustive guide to both Southern and Northern Hemisphere objects.

Peck, W. *The Constellations and How to Find Them: Southern Hemisphere.* Gall and Inglis, 1963.

Nangle, J. et al. *Stars of the Southern Heavens.* Angus and Robertson, revised edition, 1958.

Theoretical and Background knowledge:

Hoyle, F. *Astronomy.* Macdonald, 1962.

Menzel, D. *Astronomy.* Thames and Hudson, 1971.

Asimov, I. *Asimov on Astronomy.* Macdonald, 1974.

Abell, G. *Exploration of the Universe.* Rinehart and Winston, 1969.

Hoyle, F. *Galaxies, Nuclei and Quasars.* Heinemann, 1966.

Rudaux, L. and de Vaucouleurs, G. *Larousse Encyclopaedia of Astronomy.* Batchworth, 1959.

Whipple, F. *Survey of the Universe.* Prentice-Hall, 1972.
Sagan, C. and Leonard, J. *The Planets.* Life Nature Series. Time-Life, 1967.
Cousins, F. W. *The Solar System.* John Baer, 1972.
Bergamini, D. *The Universe.* Life Nature Series. Time-Life, 1964.
Evans, D. S. *Teach Yourself Astronomy.* Teach Yourself Books, third edition, 1970.
Wood, H. *Unveiling the Universe.* Angus and Robertson, 1967.

Telescope Making:

Amateur Telescope Making, Volumes 1 to 3. Scientific American, 1940, 1949, 1956. Comprise authoritative articles both originally published in the magazine and specially written for the books. Out of print and probably only accessible in the astronomical society libraries.
Muirden, J. *Beginner's Guide to Astronomical Telescope Making.* Pelham Books, 1975.
Howard, N. *Handbook for Telescope Making.* Faber and Faber, 1962.
Texereau, J. *How to Make a Telescope.* Doubleday Anchor, 1963.
Thompson, A. J. *Making Your Own Telescope.* Sky Publishing Corporation, 1947.
Page, T. and Page, L., editors. *Telescopes.* Macmillan, 1966.

Astronomical Photography:

Rackham, T. *Astronomical Photography at the Telescope.* Faber and Faber, 1959.
Paul, H. E. *Outer Space Photography for the Amateur.* Amphoto, 1960.
Keene, G. T. *Star Gazing with Telescope and Camera.* Chilton Books, 1962.

Magazine:

Sky and Telescope. Sky Publishing Corporation, Cambridge, Mass. U.S.A.
A well-illustrated monthly magazine of more than 60 pages containing numerous authoritative articles on all aspects of astronomy orientated to the amateur, and sometimes to very advanced observers or instrument makers. A special section is devoted to amateur telescope making and articles appear frequently on optics, radio astronomy, astronautics and the space programme, amateur and professional astronomers and many specialised subjects. Highly recommended.

Display Material, Visual Aids:

The Sky Publishing Corporation publishes a number of pictures, charts and other visual aids which are regularly advertised in 'Sky and Telescope' magazine. A number of other advertisers in the magazine supply slides, films, pictures and display and educational materials of all descriptions.

Appendix Eleven

GLOSSARY

ALTITUDE. Distance above horizon, measured in degrees from zero to ninety at the zenith.

APHELION. The point in an orbit furthest from the Sun. The point closest to the Sun is PERIHELION.

APOGEE. The furthest point from Earth in the orbit of the Moon or an artificial satellite. The point in an orbit closest to Earth is PERIGEE.

APPARENT POSITION (or APPARENT PLACE). The position of an object on the sphere of the sky. It does not relate to three-dimensional reality, thus, for instance, a statement that a planet will be close to the Moon on a certain date means that they will appear close to each other in the sky, not that they are actually close in space.

ASTEROID (or MINOR PLANET). One of the host of small bodies that orbit the Sun, mainly between the orbits of Mars and Jupiter.

ASTRONOMICAL UNIT (or AU). The mean radius of Earth's orbit, 149,600,000 km. A unit much used for distances within the solar system because of the relationship between planetary orbital radii established by Kepler's Third Law of Planetary Motion.

BLACK HOLE. The final state in the collapse of a star and the compression of its material into smaller volume and greater density. Nothing, not even light itself, can escape the high surface gravity of a black hole. It is detectable only through the effects of its powerful gravitation on nearby bodies.

BLUE SHIFT. see RED SHIFT.

CELESTIAL SPHERE. The sphere of the sky, viewed as a sphere of very large radius, upon which all celestial objects are deemed to be for the purpose of determining APPARENT POSITION.

CLUSTERS. Families of stars, ranging from a few to many thousands. The stars are physically associated in space, that is, they are real, not merely apparent, associations. Principally there are GALACTIC CLUSTERS (or OPEN CLUSTERS) which may contain few or many stars, are open, irregular in structure, may contain gas and dust, and may be quite young (on the galactic time scale), and GLOBULAR CLUSTERS, dense spherical aggregations of many thousands of stars, usually much older than the Sun, often strongly condensed in their central regions. Galactic clusters are found in the spiral arms of the galaxy, globular clusters in the halo, the less densely populated spherical volume.

COMETS. Bodies which show a gassy appearance, orbit the Sun, and which sometimes release gas and dust as a result of heating by the Sun during a close approach. All comets are believed to consist of a solid head, or NUCLEUS, not more than a few kilometres in diameter, probably composed of frozen material. There may or may not be a rocky core. When solar heat evaporates the frozen material, gas and dust form a head, or COMA, from which material streams away to form a TAIL.

CONJUNCTION. A planet is said to be in conjunction with another planet, the Moon or Sun when the APPARENT POSITIONS of the two bodies are close together.

CONSTELLATION. One of the eighty-eight groups into which the stars are divided for convenience in classification. All are named—the older, northern constellations by designations conferred in antiquity, the southern groups more recently. All constellation boundaries now lie on lines of RIGHT ASCENSION and DECLINATION.

DECLINATION. The celestial equivalent of latitude, measured in degrees from Celestial Equator to either Celestial Pole, designated minus in the Southern Celestial Hemisphere and positive in the Northern Celestial Hemisphere.

DEGREE. 1/360 of a circle. The apparent diameters of Moon and Sun are almost exactly one-half degree. A table of other useful measurements on the celestial sphere appears on page 81.

ECLIPSE. The Sun is said to be eclipsed when obscured by the Moon; the Moon is eclipsed when immersed in Earth's shadow. A satellite of any planet may be eclipsed in the shadow of its primary. Eclipses may be TOTAL when the Moon completely covers the Sun, or the Moon is entirely immersed in the umbral region of Earth's shadow, or PARTIAL when the Moon does not completely cover the Sun, or the Moon does not pass wholly into the UMBRA. An eclipse of the Sun can also be ANNULAR when Earth is closest to the Sun and the Moon most distant from Earth. At such times the apparent size of the Moon is slightly smaller than the apparent size of the Sun and an annulus (ring) of unobscured Sun surrounds the Moon at mid-eclipse.

ECLIPTIC. The projection of the yearly path of the Sun against the CELESTIAL SPHERE and also, therefore, the orbital path of the Earth. It is a great circle in the sky, splitting the CELESTIAL SPHERE exactly in half. The name arises from the fact that this is the plane in which ECLIPSES occur.

ELONGATION. The separation between a planet and the Sun as seen from Earth, either generally as in 'maximum (or minimum) eastern (or western) elongation', or as a precise value in degrees. Also the separation of a satellite from its planet.

EVENING STAR. Classical or poetical name given to Venus when in the evening sky (sometimes given also to Mercury and thus confusing).

GALAXY. A large system of stars, clusters, nebulae, etc. That in which the Sun is situated is called the Milky Way Galaxy. Galaxies are the basic unit of the material structure of the universe.

LIGHT-YEAR. The distance travelled by light in one year at a speed of 299,792.5 km/sec. Approximately 9,000,000,000,000 km. So also light-minute, light-hour, light-day, light-month, etc.

MAGNITUDE (APPARENT and ABSOLUTE). APPARENT MAGNITUDE is a measure of how bright a celestial object appears to us. The Sun is –27, the Moon –12 and the planets and brightest stars down to –1 and 0. Fainter magnitudes than these are positive numbers, the larger the number, the fainter the object. In good conditions the unaided eye can see to magnitude 6, binoculars and small telescopes allow us to see to about magnitude 9. ABSOLUTE MAGNITUDE is a measure of the actual brightness of the star. Thus, if all stars were at an equal distance, the brightest absolute magnitudies would also be the brightest apparent magnitudes. As the stars are at greatly varying distances, this is not so. For instance, two close stars, Alpha Centauri and Alpha Canis Majoris (Sirius), have faint absolute magnitudes, but because of their proximity appear as bright or brighter than stars of brighter absolute magnitude, such as Beta Centauri, which is much more distant. In this book 'magnitude' always means apparent magnitude, unless absolute magnitude is specifically stated.

MERIDIAN. The imaginary line passing north to south through the observer's ZENITH.

MESSIER CATALOGUE. List of 109 objects which the compiler, the French astronomer Messier (1730-1817), believed could be confused with comets. Most of these objects are star clusters and galaxies. The Messier number is used to designate these objects by astronomers today.

METEORS. Particles ranging in size from dust grains to asteroids, visible as streaks or fireballs when they enter Earth's atmosphere and become incandescent. If part of a meteor reaches the ground, it is called a METEORITE.

MORNING STAR. Venus, and sometimes Mercury, in the morning sky (see also EVENING STAR).

NEBULA. Literally meaning 'cloud-like' and applied originally to several classes of object with only this appearance in common. Now confined to clouds of gas and dust, which may be glowing or dark. Until recently, galaxies were often termed 'extra-galactic nebulae'.

NEUTRON STAR. A star less dense than a BLACK HOLE, but compressed until the planetary electrons of its atoms have collapsed onto their nuclei, forming electrically neutral objects called 'neutrons'. These, in turn, are compressed until they are in virtual contact. Most of the internal space of the original atomic structure having thus been lost, the resulting star is both very small and very dense. Long predicted in theory, but only actually identified for the first time in 1969.

NOVA. Literally 'new star', but in fact a star which has undergone a violent flaring, increasing its brightness a thousand times or more. The phase is short-lived, the star fading rapidly to its former magnitude. A far more violent event, in which a star loses much of its material to space and increases its brightness a million times or more, is termed a SUPERNOVA.

OCCULTATION. The eclipsing of a star or planet by the Moon, or by a planet.

OPPOSITION. Applied only to an outer planet, which is said to be at opposition when Sun, Earth and planet (in that order) lie on a straight line. An outer planet is best viewed when close to opposition. It then rises at sunset, is on the meridian at midnight and sets at dawn.

ORBIT. The path of an object around its primary: for instance, the path of Earth around the Sun, or the Moon around Earth.

PENUMBRA. see discussion under UMBRA.

PERIGEE. see APOGEE.

PERIHELION. see APHELION.

PHASE. The changing appearance of the Moon or a planet, depending on how much of the illuminated face is visible from Earth. Moon, Mercury and Venus show a full range of phases; Mars and Jupiter gibbous phase only; more distant planets no phases at all.

PRECESSION (as in 'precession of the equinoxes'). The axis of Earth does not point constantly in one direction but describes a small circle on the celestial sphere, taking 26,000 years to do so. Stars which lie on or close to this circle in the northern or southern sky will be pole stars at some time. Another consequence of precession is to alter the distribution of the sky between Northern and Southern Hemispheres, thus in future epochs objects too far north to be visible from New Zealand now will be visible.

PULSAR. Probably the central remnant of a star tremendously compressed by a SUPERNOVA explosion, rotating rapidly and causing short, sharp, regular pulses of radio energy by interaction of its magnetic field with surrounding gas clouds.

QUASAR. Literally 'quasi-stellar' object of star-like appearance, apparently big (of the order of one light-month across), distant (at the edge of the known universe) and an immensely powerful source of radiant energy. Quasars are probably the nuclei of galaxies in an eruptive phase.

RADIANT. The point on the celestial sphere from which the trails of shower meteors appear to diverge. The trails are in fact parallel and the effect is entirely one of perspective.

RED SHIFT. Movement of an object away from an observer causes a reddening of its light. Measurements of the shifts of lines in its spectrum (relative to a stationary comparison source) allow speed of recession to be calculated. Objects approaching an observer exhibit a BLUE SHIFT. Galaxies show an increasing red shift with distance, i.e. the most distant galaxies are moving away faster. The change in the light is detectable only by measurement of lines in photographed spectra, it is not discernible to the eye.

RIGHT ASCENSION. The celestial equivalent of longitude, sometimes given in degrees but usually in hours, minutes and seconds of SIDEREAL TIME eastwards along the Celestial Equator.

SIDEREAL TIME. The interval in hours, minutes and seconds since the last MERIDIAN PASSAGE (crossing of the meridian) of the base-point of the RIGHT ASCENSION system, the First Point of (the constellation) Aries (the Spring Equinox of the Northern Hemisphere). Local sidereal time is the RIGHT ASCENSION on the MERIDIAN.

SOLAR SYSTEM. The Sun and its family—planets, satellites, asteroids, comets, meteors—considered as a single entity.

SPECTRUM. The full range of radiation, from the shortest light waves to the longest radio waves. The visible spectrum lies between blue and red and is seen in the rainbow, or when sunlight passes through a prism.

UMBRA and PENUMBRA. Earth has a shadow, thrown by the Sun and extending far beyond the orbit of the Moon. The inner part of the shadow is total and is termed the umbra. When the Moon is in the umbra, it is totally eclipsed and an observer on the Moon would see the Sun completely obscured by Earth. The outer part of the shadow is partial and is called the penumbra. When in the penumbra, the Moon is not usually greatly dimmed and an observer on the Moon would see the Sun only partly covered by Earth.

UNIVERSAL TIME (or U.T.). Greenwich Mean Time in an astronomical context. In New Zealand, U.T. is 12 hours behind New Zealand Standard Time. Don't forget that when Daylight Saving Time is in effect you must allow for the additional correction.

VARIABLE STAR. A star which does not shine with a constant magnitude. Variation may be regular or irregular, of short or long period, minor or catastrophic.

WHITE DWARF. The first and least of the compressive stages which a star may undergo. The mass of the star has been compressed to a volume no greater than that of a planet and a few cubic centimetres of material would weigh many tonnes. Significantly greater stages of compression, in which further fundamental changes have been undergone by the matter involved, produce NEUTRON STARS and BLACK HOLES.

ZENITH. The point on the celestial sphere directly above the observer. It is an interesting and informative experiment to determine the star apparently in the zenith while standing upright, then to lie flat on the ground and see which star really is in the zenith.

ZODIAC. The constellations through which the ECLIPTIC passes. Because the orbits of the planets lie at only small angles to that of Earth, the apparent paths of the planets also pass through the Zodiacal constellations. Note that there are thirteen, not the twelve of astrology. They are: Aries, Taurus, Gemini, Cancer, Leo, Virgo, Libra, Scorpius, Ophiuchus, Sagittarius, Capricornus, Aquarius, Pisces. Ophiuchus has no sign and is the constellation ignored by astrology. Libra is the only one whose sign (the Scales) is an inanimate object.

INDEX

Page numbers in bold face indicate illustrations.